Homo-Securitas:

Business Security Evolution in the Age of Web3

Elite exclusivity

Dysart Detroit Dallas

Published by Elite Exclusivity LP
Send all Inquiries to: EliteExclusivityPR@gmail.com and Info@EliteExclusivity.com

ISBN Paperback: 978-1-952651-41-0

Contents

Introduction

"There are people in every time and every land who want to stop history in its tracks. They fear the Future, mistrust the present and invoke the security of a comfortable past, in fact, never existed."

Robert Kennedy[1]

"If you tell a lie big enough and keep repeating it, people will eventually come to believe it. The lie can be maintained only for such a time as the State can should the people from political, economic, and/or military consequences of the lie. It becomes vitally important for the State to use all of its powers to repress dissent, for the truth is the mortal enemy of the lie, and thus by extension, the truth is the greatest enemy of the state."[2] When someone posts this quote, it is a good test to see if they are ready to move forward in their life. For those who are content with the way the world is, the 9-5 job, and the accepting the drivel which spews from their television, phone, and other electronic devices, the often write this quote of as a clever phrase from *1984*[3] or *Brave New World*,[4] or if they are really edgy from *A Handmaid's Tale*.[5] However, for those who are willing to remove the blinders and look at the world for what it really is- they know the truth, that this is a

---

[1] (2024). "Security Quotes, Robert Kennedy." Brainy Quote at https://www.brainyquote.com/topics/security-quotes.
[2] Staff (2024). "Joseph Goebbels: On the Big Lie." Jewish Virtual Library at https://www.jewishvirtuallibrary.org/joseph-goebbels-on-the-quot-big-lie-quot.
[3] Orwell, George (2021). "1984." Penguin Classics: New York.
[4] Huxely, Aldous (2010). "Brave New World." Vintage: London.
[5] Atwood, Margaret (1998). "The Handmaid's Tale." Anchor Books: New York.

quote from Joseph Goebbels, an ideology that was used to maintain control over Germany during the Nazi reign of terror.[6]

In the political rhetoric today, underpaid public school educators teach that Nazi's and Socialists, are diametrically different; however, they forget that Goebbels control major portions of the Nazi party and that his branch of the National Socialist Workers Party was a rival to Hitlers in the early years of Hitler's rise to power.[7] Goebbels was a member of Stasser's northern branch of the Nazi party, which was known to have a much more socialist outlook than the primary party.[8] In 1926, the northern party started to revise the southern party's positions.[9] Ironically, bitter enemies as they were, Stalin used Goebbels ideology in the control of the Soviet Union- maintaining a fiction which kept people under the thumb of the repressive government.

So, what does this have to do with digital security? In the digital world, everything is data. Data controls your wealth. Data controls your schedule. Data controls what you know and what you do not know. For humanity to advance past the current dark age, people need to have access to data. Just like when Gutenberg created the printing press, which led to people having access to immutable writings (books), the advent of the blockchain has created a system of immutable digital data. While it is commonly seen as the difference between Web 2.0 and Web3 is centralization v. decentralization- the immutability of the blockchain is also a paradigm shift.[10] Data stored on the chain cannot be

---

[6] JWL *supra.*
[7] *Manvell, Roger; Fraenkel, Heinrich (2010) [1960]. Doctor Goebbels: His Life and Death. New York: Skyhorse.*
[8] *Id.*
[9] *Reuth, Ralf Georg (1994). Goebbels. Harvest.*

destroyed- only buried.[11] This means that data security is the lifeblood of freedom in the new era and there are those, who like Goebbels in 1926 wanted to control information, who wish to hold back human advancement.

In the prelude, the Kennedy quote looks at who is the enemy of progress. For most of us, we see the myth of the lazy person holding back human progress. This is a useful myth for those who want to stop human progress and keep things "the way they are." For those who are profiting off of exploitation of the ignorant (which is why we need free access to data), keeping a scapegoat is important. For conservatives who hate progress, the scapegoat is "illegal aliens." For liberals who hate progress, it is "conspiracy theories." For libertarians who hate progress, the scapegoat is "the government." For progressive who hate progress, the true enemy is "anyone who wants to protect data and ensure that it is not stolen, buried, or corrupted." If you are reading this, then I guess you are in that boat with the rest of us, you want your data to be secured and you want to make sure that it is correct.

One of the greatest talking heads representing the "anti-crypto" crowd is Elizabeth Warren, the senior Senator from Massachusetts. In January 2023, she stated in an interview, "For all their talk of innovation and financial inclusion, the crypto industry giants- from FTX to Celsius to Voyager- are collapsing under the weight of their own fraud, deceit,

---

[10] Bajaj, Aryan (2022). "What's the Difference between Web 2.0 and Web 3.0?" Analytics Vidhya at https://www.analyticsvidhya.com/blog/2022/06/whats-the-difference-between-web-2-0-and-web-3-0/.
[11] Staff (2024). "What is Immutable Ledger in Blockchain and Its Benefits?" Solu Labs at https://www.solulab.com/what-is-immutable-ledger-in-blockchain-and-its-benefits/.

and gross mismanagement."[12] Lets break this down piece by piece:

- FTX- The FTX collapse was caused by lack of oversight. Alameda Research was funding risky investment to the point that the reserve of company was no longer able to cover its liabilities.[13] The reason this error was caught was because the crypto news agency *CoinDesk* reported on the poor financial health of Alameda, causing $5 Billion to be withdrawn in one day. As FTX was an exchange operating in the United States- the United States government failed to uphold its duty under present commerce law to ensure that FTX's reports were true.[14] Years of negligence by the federal government, of the current laws, created an imbalance which cost people millions of dollars. The crypto industry policed itself- revealing the failure of the government to enforce current laws.
- Celsius- Celsius had risky investment, with yields of up to 17%. This was an odd way to invest money, but they let people know the risk.[15] When the Terra/Luna attack occurred in 2022, the lost of value for the Terra ecosystem, which was a major

---

[12] Morrow, Allison (2023). "Elizabeth Warren: Crypto giants are collapsing under the weight of their own fraud." CNN.com at https://www.cnn.com/2023/01/25/investing/crypto-elizabeth-warren-ftx/index.html.

[13] Reiff, Nathan (2023). "The Collapse of FTX: What Went Wrong with the Crypto Exchange?" Investopedia at https://www.investopedia.com/what-went-wrong-with-ftx-6828447.

[14] *Id.*

[15] Biswas, Soma (2023). "Celsius Network, One of Crypto's Biggest Collapses Ends with Bankruptcy Case." Wall Street Journal Pro at https://www.wsj.com/articles/celsius-network-one-of-cryptos-biggest-collapses-ends-bankruptcy-case-85bb7efd.

investment vehicle of Celsius, cause a loss in value. The lack of clarity from Celsius on how they were dealing with the problem led to crypto news agencies reporting on the matter and the loss of confidence.[16] In June, Celsius filed for bankruptcy under the law. Once again, the government did not maintain common securities practices, including allowing firms to short major investment's in Terra while meanwhile shorting the medium used to make the short- which is a form of market manipulation expressly forbidden.[17] Rather than punish the bad actors, the government went after the victim- the owner of Terra.

- Voyager- Voyager loaned $650 Million in assets to 3 Arrows Capital, a hedge fund.[18] 3AC defaulted resulting in a major impact to voyager's liquidity. Reports surfaced that Voyager was not complying with regulations, yet government officials were "building a case" against them.[19] Oddly, few congresspeople or government officials, many of whom are heavily invested in the cryptocurrency

---

[16] Napolitano, Elizabeth (2022). "The Fall of Celsius Network: A Timeline of the Crypto Lender's Decent into Insolvency." CoinDesk at https://www.coindesk.com/markets/2022/07/15/the-fall-of-celsius-network-a-timeline-of-the-crypto-lenders-descent-into-insolvency/.

[17] Biswas, *supra.*

[18] Nelson, Danny & Morris, David (2022). "Behind Voyager's Fall: Crypto Broker Acted Like a Bank, Went Bankrupt." CoinDesk at https://www.coindesk.com/layer2/2022/07/12/behind-voyagers-fall-crypto-broker-acted-like-a-bank-went-bankrupt/.

[19] FTC Website at https://www.ftc.gov/news-events/news/press-releases/2023/10/ftc-reaches-settlement-crypto-company-voyager-digital-charges-former-executive-falsely-claiming, This also brings into question the United States Federal Government using the civil law system like a private entity to avoid the standard of proof listed in the constitution.

> market, lost money.[20] This not only shows poor acting on the part of the government, allowing people to be exploited for the important case success rate, but also information links which allowed insiders to make money on the Voyager loss.

Mismanagement by the government in all three cases allowed for these disasters to happen- however, talking heads like Warren blame the entire crypto community. The question is why Sam Bankman Fried was speaking to congress on cryptocurrency regulation after his arrest, rather than the thousands of legitimate crypto company owners who are trying to comply with the quagmire of laws offered by the federal government.[21]

But Liz Warren's intent does not end there. She attacks the past making martyrs of the corrupt and villains of the victims, but she also attacks the present.[22] Those who are willing to work with cryptocurrency, such as Ro Khanna (D-Cal) and French Hill (R-Ark), are common opponents of hers as she attacks those who are willing to work with the industry to ensure compliance and clear laws. Activists such as Warren need confusion to keep people circling, following more of Gobbel's and Hitler's teachings, never let the hate die down.[23]

---

[20] Nelson & Morris, *supra*.
[21] Banks, David (2022). "Crypto was meant to solve financial corruption. The FTX Scandal shows it's got worse." The Guardian at https://www.theguardian.com/commentisfree/2022/dec/15/crypto-financial-and-corruption-making-it-worse-ftx.
[22] Her Own Website, at https://www.warren.senate.gov/newsroom/press-releases/warren-demands-answers-from-crypto-firms-on-abuse-of-the-revolving-door-to-undermine-bipartisan-efforts-to-rein-in-crypto-terrorist-financing.
[23] Reuth, *supra*.

Perhaps most alarming is the attack we are seeing on the future of data by the anti-cryptocurrency crowd. This is drawing all of the digital industry into a whirlpool which is centered around the pit of lies and deception in Washington D.C. Anyone who understands business knows that access to information and the reliability of that information is paramount in making decisions.[24] To this end, the attacks against the blockchain (as for many it is synonymous with crypto) are attacking the entire digital field.[25] In many cases, those who attack the industry conflate cryptocurrency with CBDC, blockchain with land based servers, the cloud with canon, and the internet with magic. They do not know how these things work- so to them anything that is new is evil, and they whip their followers into a frenzy in order to undermine what they do not understand.[26]

This book is designed to go over the areas of security needed for a business to function in the world which we have created. Especially in the United States, we elect these idiots that are making our lives harder. This means that we, and our IT departments need to work or smarter harder to get good results. One of the best ways to work smarter rather than harder is to have good security. This keeps your company out of the headlines (thus out of the crosshairs of regulators and watchdogs), it keeps your customers happy

---

[24] Encila, Christian (2024). "Blockchain Group Declares 'War' -US Senator's Anti-Crypto Bill Endangers National Security." Bitcoinist at https://bitcoinist.com/blockchain-group-declares-war-vs-anti-crypto-bill/.

[25] *Id.*

[26] Reynolds, Sam (2023). "Bipartisan Anti-Crypto Terror Financing Bill Heads to U.S. Senate." CoinDesk at https://www.coindesk.com/policy/2023/12/08/bipartisan-anti-crypto-terror-financing-bill-heads-to-us-senate/.

(thus out of their congresspeople's offices), and it keeps your business running smoothly. As the future is data security in Web3, your company has to be ready for this future.

Methodology

While a pretty book looks nice on a shelf, this book is a manual for the war on those who are after your data. Ten years ago, this would have been faceless hacker groups who were just looking for a quick buck exploiting data problems. Now this group encompasses state actors, political activist groups, corporate spies, and a whole host of bad actors- which still includes the hacking groups. This means that digital security and general security for a company needs to be part of the fabric of our business DNA.

The book itself is taking a meta-synthesis of what is available on the market for good digital security. While it cannot be understated that a good security consultant can help you with the fine points in your security system, this book looks at the broad strokes that can function for almost any kind of business. Every business has:

1.) Employees;
2.) Employees with Access to Digital Data;
3.) Problems that Crop Up;
4.) The potential to have attacks against them;
5.) Physical Security Needs;
6.) Issues with Digital Identity and Intellectual Property;
7.) To Deal with the Advent of AI;
8.) To Deal with the Fact the World is More Interconnected.

In each part of the book, we address one of these factors. This allows the book to be used like a handbook after your first readthrough.

Part I deals with data privacy and security. Regardless of what business your company is in, it has employees. Employees deal with data in some shape or form. This means that you need to ensure that your employees are handling any data that they have access to securely. This is the first and most basic step of any security plan is to make sure that your own people are working toward the same goal to which you are working.

Part II deals with account security. This goes beyond the simple "employees know things" to accounts where multiple people have access. With each additional person who has access to a system, the system is less secure. By planning for this change in security, you have the ability to scale the security with the levels of access and delineate who has access to what. This allows for a more controlled system for your company.

Part III deals with attacks. If your company has money, then there is a chance that someone will choose to attack your company to take it. If your company does not have money, then there is the chance that someone will try to exploit your company and force you to come up with money to keep from being crushed under a scam. The key to beating these attacks is to ensure that your employees know what to look for and know how to keep their portals secure.

Physical security is another important part of a security plan for a digital company. Regardless of whether you are working from a home office or from an office building, you still need to ensure that your hardware and files are security. This means that you need to have a physical security plan. This can be as simple as good locks to as complex as guards on randomized rotations. This also includes ensuring that your trash is disposed of properly.

Part V looks at who you are, not who you are as a person, but who you are in the digital sense. As KYC and AML become important factors in the world, businesses need to make sure that their business identity is not taken away from them digitally. Scammers know that people will open emails and letters from certain companies, this means the more popular you are the more likely you are to come under attack or be used in an attack against someone else. Control of your digital identity is paramount in ensuring that your system is protected.

Part VI looks at your computer system. While your employees can be trained in all sorts of ways to provide for security; if your systems have problems- then you have problems. Algorithmic bias and discrimination can weaken a company and cause dissent within the workforce. Analyzing this can be difficult, but it is worthwhile.

Next, we talk about the use of AI and Machine Learning in your security. While this field is still young, there is the ability to ensure that someone is on guard at all times. This allows you to augment your human security in a way that humans could never do. Having strong AI and ensuring that it is up to date protect not only your digital environment but also your physical plant as well.

Finally, Part VIII acknowledges that no matter what type of business you have, you are now working on the global level. Whether you are selling pots in Kyoto Japan or doing plumbing in Altoona, Pennsylvania; if you are building a hedge fund in London, or selling bonds in Berlin- you are working on a global stage. Entering business with this mindset of work local, defend global can help your business have a strong plan.

Methodologically, this book also uses AI editing. While the raw text was created by the author, it has been edited by multiple AI systems. This allows for the text to have limited, if any, errors to make the reading easier for the reader. It is a known fact that some AIs like some words, so we apologize if words like "delve" and "tapestry" are overused in the book- we have tried to removed them from the edited work as much as possible.[27]

This book is apolitical, as far as the classic battle between conservatives and liberals is concerned. The paradigm has shifted between classic conservatives and classic liberals to those who want to see humanity advance and those who wish to see it stagnate. I hope that you as a reader are in the latter group. This book is about ensuring progress by ensuring security within systems.

---

[27] The process for this book is write the book, have AI edit the chapters, piece by piece, go back through and edit the work edited by the AI to ensure that the AI did not change the meaning of any of the passages.

Part I: Data Privacy and Security

Introduction to Part I: Establishing Foundational Security Principles

"If you want total security, go to prison. There you're fed, clothed, given medical care and so on. The only thing lacking… is freedom."

Dwight D. Eisenhower[28]

In the dynamic landscape of the Web3 era, where digital transformation and technological advancements have revolutionized the way organizations operate, the importance of robust security measures cannot be overstated. Part I of this book delves into the foundational principles and strategies essential for establishing a secure operational environment within organizations. From securing individual workstations to implementing robust access control measures, each chapter in this section provides valuable insights and actionable guidance aimed at fortifying defenses and mitigating risks.

The chapters in Part I are designed to equip readers with a comprehensive understanding of the fundamental elements of cybersecurity and physical security, laying the groundwork for building a resilient security posture in the digital age. As organizations increasingly rely on digital technologies and interconnected systems to drive innovation and streamline operations, the need to safeguard critical assets and sensitive information from cyber threats and physical intrusions has never been more pressing.

At the core of Part I is the recognition that an individual's computer serves as their fortress in the digital realm. The

---

[28] (2024). "Security Quotes, Dwight D. Eisenhower." Brainy Quote at https://www.brainyquote.com/topics/security-quotes.

first chapters emphasize the importance of implementing stringent login procedures, access controls, and monitoring systems to protect against unauthorized access and potential vulnerabilities. By instilling a culture of security awareness and accountability among employees, organizations can fortify their defenses and minimize the risk of data breaches or cyberattacks. Furthermore, the exploration of password policies underscores the critical role of authentication mechanisms in safeguarding sensitive information from unauthorized access. Through the adoption of advanced encryption techniques and password management practices, organizations can enhance the security posture of their digital infrastructure and mitigate the risk of password-related vulnerabilities.

In addition to digital security, Part I also addresses the importance of securing physical assets and locations. Chapters on data storage device options, backup systems, and access control mechanisms highlight the need for comprehensive security measures to protect against data loss, theft, and unauthorized access. By implementing encryption technologies, access controls, and robust backup strategies, organizations can ensure the integrity and availability of their data assets in the face of evolving threats. The discussion on access control and perimeter security underscores the significance of implementing layered security protocols to safeguard physical locations and assets. By deploying surveillance systems, access badges, and visitor management procedures, organizations can mitigate the risk of unauthorized access, tailgating incidents, and insider threats, thereby enhancing overall security posture.

Part I serves as a primary guide for organizations seeking to establish a secure operational environment in the Web3 era. By embracing the principles and strategies outlined in this section, organizations can build a resilient security foundation, protect their digital and physical assets, and effectively mitigate the evolving threats posed by cybercriminals and malicious actors. As we delve deeper into subsequent sections, we will explore advanced concepts and emerging technologies aimed at further enhancing security in the digital age.

Chapter 1: A Person's Computer is Their Fortress

"I Think its fair to say that personal computers have become the most empowering tool we've ever created. They're tools of communication, they're tools of creativity, and they be shaped by their user."

Bill Gates29

This chapter addresses critical aspects of cybersecurity within organizational settings, focusing on various measures to safeguard sensitive information, prevent unauthorized access, and ensure compliance with security protocols. Here we work from the concept that a person's computer is their fortress, their fiefdom in the preservation and protection of the company. To this end, a company has a duty to empower their employees with the ability to maintain the protect the corporate realm through proper management and maintenance of their computer and access.

The first section emphasizes the importance of granting employees access to their own computers while maintaining strict control and oversight to uphold organizational security standards. Implementing measures such as login/logout procedures, time restrictions, and website access restrictions helps mitigate security risks and prevent unauthorized activities. Next, the chapter highlights the pivotal role of organizations in safeguarding access to employees' workspaces, emphasizing the need for robust security measures to identify and prevent potential threats before they occur. By implementing stringent access

[29] (2024). "Personal Computer Quotes: Bill Gates." AZ Quotes at https://www.azquotes.com/quotes/topics/personal-computers.html.

controls and monitoring mechanisms, organizations can proactively mitigate security risks and protect sensitive information.

Furthermore, the chapter discusses the importance of restricting IT access to backups to prevent unauthorized access to sensitive data and ensure data integrity and confidentiality. By limiting IT personnel's access to backups and implementing strict access controls, organizations can mitigate the risk of data breaches and unauthorized access to critical information. Another crucial aspect covered in this chapter is the implementation of restricted access systems to protect sensitive networks and systems from unauthorized access. By limiting access to authorized personnel only and implementing stringent access controls, organizations can enhance security and prevent unauthorized access to critical systems.

Lastly, the chapter explores the use of monitoring software as a security measure and the associated risks and challenges. While monitoring software plays a crucial role in enhancing security and compliance enforcement, organizations must address privacy concerns, ensure proper training for personnel responsible for monitoring, and mitigate the risk of false positives or misinterpretation of data. By implementing robust security measures, restricting access to sensitive information and systems, and carefully monitoring and managing access controls, organizations can effectively enhance cybersecurity posture, protect sensitive information, and ensure compliance with organizational policies and regulatory requirements.

Access to One's Own Computer

When talking about Web3 security, the fundamental premise is that your computer is your fortress. However, it's crucial to understand that this fortress, while in your possession, ultimately belongs to the company you work for.[30] Therefore, accessing your computer is not just a matter of convenience; it's a privilege granted by the company to carry out your work responsibilities. As such, strict measures must be in place to ensure that this privilege is not abused and that the company's assets remain protected at all times.[31]

First and foremost, access to your computer should be restricted to authorized personnel only.[32] This means implementing robust authentication mechanisms such as username and password combinations, biometric scans, or multi-factor authentication. Logging in should require providing verifiable credentials that uniquely identify you as an authorized user of the company's system. Additionally, logging out should be mandatory whenever the computer is left unattended, even for brief periods, to prevent unauthorized access by others.

Furthermore, access to specific resources and websites should be regulated based on job roles and responsibilities.[33] For instance, employees in finance may require access to financial data and accounting software, while those in marketing may need access to social media platforms and analytics tools. Implementing role-based access controls ensures that each employee has access only to the resources

---

[30] *See* Computer Fraud and Abuse Act (1986).
[31] *See Van Buren v. USA*, 593 U.S. ____ (2021).
[32] Citation for Sidebar: Arios96 (2022).
"IT Pranks for users that leave their computer Unattended." Spiceworks at https://community.spiceworks.com/topic/2345845-it-pranks-for-users-that-leave-their-computers-unattended.
[33] *See* CFAA *supra.*

necessary for performing their job functions, minimizing the risk of unauthorized data exposure or misuse.

Time-based restrictions can also be instrumental in enhancing security.[34] By defining specific time windows during which certain activities or access privileges are allowed, companies can mitigate the risk of unauthorized access outside of business hours. For example, restricting access to sensitive systems or data repositories during non-working hours can significantly reduce the likelihood of data breaches or malicious activities occurring outside of regular business operations.[35]

Moreover, monitoring and logging user activities are essential components of maintaining security and accountability.[36] By logging login and logout times, as well as recording access attempts and usage patterns, companies can track employee behavior and identify any anomalies or suspicious activities that may indicate a security threat. Real-time monitoring tools can alert security personnel to potential security incidents, allowing for immediate response and mitigation measures.

In summary, treating your computer as your fortress means recognizing that it is a vital asset of the company and must be protected accordingly. Access should be restricted to authorized personnel, with stringent authentication mechanisms, role-based access controls, time-based restrictions, and comprehensive monitoring in place to safeguard company data and assets. By adhering to these

---

[34] McCarthy, Maile (2023). "Principle of Least Privilege Explained." Strong DM at https://www.strongdm.com/blog/principle-of-least-privilege.
[35] *Id.*
[36] *See* services such as WorkTime at https://www.worktime.com/login-logout-monitoring-worktime.

principles, employees can play an active role in maintaining the security of their organization's digital infrastructure.

Access to Workspace

A company's policies play a pivotal role in safeguarding access to individuals' workspaces. Recognizing that employees are the frontline defenders against cyber threats, it is incumbent upon organizations to provide the necessary tools and infrastructure to fortify their digital perimeters. The more protection employees have within their workspace, the better equipped they are to identify and mitigate potential threats before they materialize into security breaches.

One of the primary responsibilities of companies is to establish robust access controls and authentication mechanisms to prevent unauthorized access to employees' workspaces. This includes implementing strong password policies, multi-factor authentication, and encryption protocols to ensure that only authorized users can gain access to sensitive data and resources.[37] By enforcing strict access controls, companies can significantly reduce the risk of unauthorized access by malicious actors seeking to exploit vulnerabilities in the system.

Furthermore, companies must invest in continuous monitoring and surveillance technologies to detect and respond to security threats in real-time. By deploying intrusion detection systems, security information and event management (SIEM) tools, and endpoint detection and

---

[37] Staff (2024). "What is Multi-Factor Authentication (MFA) and How Does it Work?" OneLogin at https://www.onelogin.com/learn/what-is-mfa.

response (EDR) solutions, organizations can proactively identify anomalous behavior and potential security incidents within employees' workspaces.[38] Timely detection allows security teams to investigate and remediate threats before they escalate into full-blown cyber-attacks.

Companies should prioritize employee education and awareness programs to enhance their cybersecurity posture. Training employees on best practices for password management, phishing awareness, and social engineering tactics empowers them to recognize and respond effectively to potential threats within their workspace. By fostering a culture of security awareness and vigilance, companies can leverage their workforce as a first line of defense against cyber threats.

Organizations should implement robust endpoint security measures to protect devices within employees' workspaces from malware, ransomware, and other malicious software. This includes deploying antivirus software, firewalls, and endpoint detection and response (EDR) solutions to detect and block malicious activity on endpoints such as laptops, desktops, and mobile devices.[39] By securing endpoints, companies can minimize the risk of data breaches and unauthorized access to sensitive information.

Company policy plays a critical role in protecting access to individuals' workspaces, and the more protection they

---

[38] Staff (2024). "Security Information and Event Management (SIEM)" Impervia at https://www.imperva.com/learn/application-security/siem/; *See Also* Staff (2024). "Endpoint Detection and Response Tools: How they Work and Solutions to Know." Cynet at https://www.cynet.com/endpoint-protection-and-edr/top-6-edr-tools-compared/.
[39] Cynet, *supra*.

provide, the better equipped employees are to identify and respond to security threats. By implementing robust access controls, continuous monitoring, employee education programs, and endpoint security measures, organizations can fortify their digital perimeters and mitigate the risk of cyber-attacks. Investing in workspace protection is essential for safeguarding sensitive data, preserving business continuity, and maintaining the trust of customers and stakeholders in an increasingly interconnected and digital world.

IT Access to Backups

Ensuring that IT personnel have restricted access to sensitive data is paramount in maintaining the security and integrity of an organization's digital assets. While IT staff require access to workstations to perform their duties effectively, it is imperative to implement measures that restrict their access to sensitive data unless absolutely necessary.[40] By striking a balance between access and restriction, companies can mitigate the risk of data breaches and unauthorized access by IT personnel.

One approach to restricting IT access to sensitive data is through the principle of least privilege, which grants users only the minimum level of access required to perform their job functions.[41] For IT staff, this means providing them with access permissions tailored to their specific roles and responsibilities, ensuring that they can perform essential tasks without unnecessary access to sensitive data. By limiting access rights based on job function and necessity,

---

[40] Staff (2023). "Best Practices for Secure Exchange of Sensitive Data with Clients for Web Agencies." Hypervault at https://hypervault.com/exchange-of-sensitive-data-agencies-best-practices/.
[41] *Id.*

companies can reduce the likelihood of accidental or intentional misuse of privileged credentials by IT personnel.

Companies can implement robust access controls and auditing mechanisms to monitor and track IT access to sensitive data.[42] This includes logging all access attempts, enforcing strict authentication requirements, and conducting regular audits of IT activities to detect and investigate any unauthorized access or suspicious behavior. By maintaining a comprehensive audit trail of IT activities, organizations can ensure accountability and transparency in their data access policies.

Corporations should consider implementing technical controls such as encryption and data masking to protect sensitive data from unauthorized access, even by IT personnel.[43] By encrypting sensitive data at rest and in transit, organizations can ensure that only authorized individuals with the necessary decryption keys can access the information. Similarly, data masking techniques can be employed to obscure sensitive data elements from IT staff during routine system maintenance or troubleshooting activities, reducing the risk of inadvertent exposure.

Restricting IT access to sensitive data is essential for safeguarding an organization's digital assets and mitigating the risk of data breaches. By adopting a least privilege approach, implementing robust access controls and auditing mechanisms, and leveraging technical controls such as encryption and data masking, companies can strike a

---

[42] *See* Data Guard Store at https://www.dataguardstore.com/Sensitive-Data-Auditing.asp.
[43] *See* Oracle Help Center at https://docs.oracle.com/en/cloud/saas/human-resources/24a/ochus/sensitive-data-access-audit.html#s20079263

balance between providing IT staff with the access they need to perform their job duties and protecting sensitive data from unauthorized access or misuse. Prioritizing data security in IT access policies is critical for maintaining trust, compliance, and resilience in an increasingly complex and interconnected digital landscape.

Restricted Access Systems

Securing physical locations where sensitive systems are housed is a critical component of an organization's overall security posture. By implementing stringent access controls and limiting access to authorized personnel only, companies can prevent unauthorized individuals from compromising sensitive systems and data.[44] Additionally, restricting access to systems with sensitive networks to only those who require access helps minimize the risk of insider threats and unauthorized access.

One approach to improving security is through the use of access control measures such as keycard or biometric authentication systems. By requiring employees to authenticate themselves using a keycard or biometric identifier before gaining access to sensitive areas, companies can ensure that only authorized individuals are allowed entry.[45] Access control systems can also be integrated with monitoring and alerting mechanisms to detect and respond to any unauthorized attempts to access restricted areas.

Furthermore, companies can implement physical barriers such as locked doors, security gates, and surveillance

---

[44] Gelmis, Ibrahim (2023). "9 Access Control Best Practices." Resmo at https://www.resmo.com/blog/access-control-best-practices.
[45] *Id.*

cameras to prevent unauthorized access to sensitive locations.[46] Physical barriers serve as deterrents to potential intruders and provide an additional layer of security against unauthorized access attempts. Regular security audits and inspections can also help identify and address any vulnerabilities or weaknesses in physical security measures.

In addition to securing physical locations, companies should also restrict access to systems with sensitive networks to only those employees who require access to perform their job duties. This can be achieved through the use of role-based access controls (RBAC), which grant users access privileges based on their specific roles and responsibilities within the organization.[47] By assigning access permissions on a need-to-know basis, companies can minimize the risk of unauthorized access to sensitive networks and data.

Ideally, companies should regularly review and update access permissions and privileges to ensure that only authorized individuals have access to sensitive systems and data. This includes promptly revoking access for employees who no longer require access due to changes in their roles or employment status. By maintaining a proactive approach to access management, companies can reduce the likelihood of security breaches and unauthorized access incidents.

Restricting access to physical locations where sensitive systems are housed and limiting access to systems with sensitive networks are essential measures for improving security and protecting sensitive data. By implementing

---

[46] Staff (2024). "What is Physical Access Control." Avigilon at https://www.avigilon.com/blog/physical-access-control.
[47] *Id.*

access control measures, physical barriers, and role-based access controls, companies can prevent unauthorized individuals from compromising sensitive assets and mitigate the risk of insider threats.[48] Prioritizing physical security alongside cybersecurity measures helps create a comprehensive defense strategy that safeguards an organization's most valuable assets from potential threats and vulnerabilities.

Monitoring Software

While monitoring software can be an effective tool for enforcing security policies and preventing unauthorized access to sensitive data, it also presents certain risks and challenges that organizations must address. One of the primary concerns with monitoring software is the potential for abuse or misuse by those responsible for overseeing the system.[49] While the software itself may prevent employees from accessing restricted areas or engaging in unauthorized activities, the individuals who monitor the software and have access to the data it collects must also be vetted and trusted.

One risk associated with monitoring software is the potential for privacy violations and employee surveillance. Employees may feel uncomfortable knowing that their actions are being constantly monitored and recorded, leading to concerns about privacy and autonomy in the workplace.[50] Additionally, the data collected by monitoring software may inadvertently capture sensitive or personal

---

[48] Gelmis, *supra.*
[49] *See* Staff (2024). "Infrastructure Monitoring Tools, Reviews and Ratings." Gartner Peer Insights at https://www.gartner.com/reviews/market/infrastructure-monitoring-tools.
[50] *Id.*

information about employees, raising questions about data protection and compliance with privacy regulations.

Furthermore, the individuals responsible for monitoring the software and analyzing the data must be carefully vetted and trained to handle sensitive information responsibly. This includes ensuring that employees tasked with monitoring software undergo background checks, receive adequate training on privacy laws and regulations, and adhere to strict protocols for handling and protecting sensitive data.[51] Failure to properly vet and train employees who have access to monitoring software can increase the risk of insider threats, data breaches, and unauthorized access to sensitive information.

Another risk associated with monitoring software is the potential for false positives or misinterpretation of data, which can lead to unnecessary alerts, false accusations, and disruptions in the workplace. Monitoring software may generate alerts or notifications based on predefined rules or thresholds, but these alerts must be carefully reviewed and verified by qualified personnel to avoid false alarms and unnecessary investigations. Additionally, the interpretation of monitoring data requires a nuanced understanding of employee behavior and context, which may be challenging for individuals who lack proper training or experience in security analysis.

While monitoring software can be a valuable tool for enhancing security and enforcing compliance with organizational policies, it also poses certain risks and challenges that must be addressed. Organizations must carefully consider the privacy implications of monitoring software and ensure that appropriate safeguards are in place

---

[5151] *Id.*

to protect employee privacy and data confidentiality. Additionally, employees tasked with monitoring software must be carefully vetted, trained, and supervised to prevent misuse or abuse of the system and to ensure accurate interpretation of monitoring data. By carefully balancing the benefits and risks of monitoring software, organizations can effectively mitigate security threats while respecting the privacy and rights of their employees.

Conclusion

This chapter lays the foundation for organizational cybersecurity by emphasizing the critical importance of safeguarding organizational assets against unauthorized access and potential security threats. The first section underscores the significance of providing employees with access to their own computers while maintaining stringent control measures to uphold organizational security standards. By implementing robust login/logout procedures, time restrictions, and website access controls, organizations can effectively mitigate security risks and prevent unauthorized activities.

Next, the chapter delves into the pivotal role of organizations in protecting access to employees' workspaces, emphasizing the need for proactive security measures to identify and prevent potential threats. By implementing strict access controls and monitoring mechanisms, organizations can bolster their security posture and safeguard sensitive information. Continuing, the chapter explores the importance of restricting IT access to backups to ensure data integrity and confidentiality. By limiting IT personnel's access to backups and enforcing strict access controls, organizations can mitigate the risk of data breaches and unauthorized access to critical information.

Another critical aspect covered in this chapter is the implementation of restricted access systems to safeguard sensitive networks and systems from unauthorized access. By restricting access to authorized personnel only and implementing stringent access controls, organizations can enhance security and prevent unauthorized access to critical systems. Finally, we discusses the use of monitoring software as a security measure and the associated risks and challenges. While monitoring software is essential for enhancing security and compliance enforcement, organizations must address privacy concerns and ensure proper training for personnel responsible for monitoring.

In summary, this chapter underscores the importance of implementing robust security measures, restricting access to sensitive information and systems, and carefully monitoring and managing access controls. By doing so, organizations can effectively enhance their cybersecurity posture, protect sensitive assets, and ensure compliance with organizational policies and regulatory requirements.

Chapter 2: Password Policies

"Treat your password like your toothbrush. Don't let anybody else use it and get a new one every six months."

Clifford Stoll[52]

This chapter investigates the critical importance of password policies in safeguarding sensitive information and fortifying cybersecurity defenses within organizations. Beginning with an exploration of the evolution of passwords, the chapter underscores the necessity for continual adaptation to combat ever-evolving cyber threats. From rudimentary practices in the 90s to modern-day sophisticated measures, the evolution of password security highlights the ongoing battle against cyber adversaries and the imperative for robust defense mechanisms.

Central to the discussion is the concept of 256-bit security, which exemplifies the stringent encryption standards employed by organizations to protect their data from unauthorized access and exploitation. This high level of encryption underscores the depth of security measures necessary to safeguard against sophisticated cyberattacks in today's digital landscape. By adopting encryption methodologies with such high cryptographic strength, organizations can significantly enhance their resilience to cyber threats.

The chapter explores the key components of effective password policies, emphasizing the need for complexity, strength, and regular changes. A robust password policy framework encompasses various factors, including minimum length requirements, character diversity, and

[52] (2024). "Password Quotes, Clifford Stoll" AZ Quotes at https://www.azquotes.com/quotes/topics/passwords.html.

prohibition of common or easily guessable passwords. Furthermore, the implementation of multifactor authentication adds an additional layer of security, further fortifying access controls and mitigating the risk of unauthorized access.

Moreover, we discusses best practices for password change policies, highlighting the importance of striking a balance between security requirements and user convenience. Clear communication, user education, and adequate support mechanisms play pivotal roles in promoting awareness and adherence to password policies among employees. By fostering a culture of password hygiene and security awareness, organizations can empower their workforce to actively contribute to the protection of sensitive data and the prevention of security breaches.

This chapter underscores the critical role of password policies in strengthening cybersecurity defenses and mitigating the risk of data breaches. By embracing robust encryption standards, implementing effective password policies, and promoting security awareness among employees, organizations can bolster their resilience to cyber threats and safeguard their valuable assets in an increasingly digitized world.

The Age of SEX and GOD

In the nascent days of the internet, passwords were often treated with a level of casualness that is unimaginable in today's security-conscious environment. The 90s heralded the Age of "SEX" and "GOD," where passwords were commonly chosen based on easily memorable words or phrases such as "sex," "god," or even "password" itself. At that time, cybersecurity was not a top priority for most users, and the concept of cyber threats was still relatively novel. As a result, individuals tended to opt for convenience over security when selecting passwords, leading to widespread vulnerabilities in online accounts.

However, as technology advanced and cyber threats became more sophisticated, the shortcomings of these simplistic passwords became increasingly apparent. Hackers capitalized on the predictability of common passwords, exploiting vulnerabilities to gain unauthorized access to sensitive information. High-profile security breaches highlighted the urgent need for stronger password policies to protect against unauthorized access and data breaches.[53]

In response to these evolving threats, the concept of password security underwent a significant paradigm shift. The era of "SEX" and "GOD" passwords gave way to a new era characterized by stringent password requirements and complex encryption algorithms.[54] The focus shifted from convenience to complexity, with users encouraged to create passwords that were long, unique, and resistant to brute-force attacks.

Today, password security remains a critical aspect of cybersecurity practices, with organizations implementing robust password policies to safeguard sensitive data. Modern password policies typically enforce minimum length requirements, character diversity, and regular password changes to enhance security measures.[55]

---

[53] Staff (2019). "4of 5 Company Breaches due to Poor Passwords." Kubera at https://www.kuberapayments.com/blog/4-of-5-company-breaches-due-to-poor-passwords.

[54] Shelton, Jacob (2023). "If your password is on this list, please rethink your life." Ranker at https://www.ranker.com/list/common-passwords/jacob-shelton/

[55] Gilbertson, Scott (2023). "The Best Password Managers to Secure

Additionally, the advent of multi-factor authentication (MFA) has introduced an additional layer of security, further mitigating the risk of unauthorized access.

In essence, the Age of "SEX" and "GOD" passwords served as a stark reminder of the inherent vulnerabilities associated with lax password practices. The evolution of password security underscores the ongoing arms race between cyber attackers and defenders, highlighting the importance of continually adapting security measures to counter emerging threats.[56] As technology continues to advance, it is imperative that organizations remain vigilant in implementing robust password policies to protect against evolving cybersecurity risks.

Concept of 256 Bit Security

The concept of 256-bit security refers to the level of cryptographic strength used to protect sensitive information, such as passwords, encryption keys, or digital signatures.[57] In cryptography, security strength is measured in bits, with higher bit lengths indicating stronger levels of security. A 256-bit key, for example, is exponentially more secure than a 128-bit key, as it offers a vastly larger number of possible combinations.

At its core, 256-bit security relies on the use of cryptographic algorithms that employ keys with a length of 256 bits. These algorithms, such as AES (Advanced Encryption Standard) and SHA-256 (Secure Hash

---

Your Digital Life." Wired at https://www.wired.com/story/best-password-managers/.

[56] Kubera, *supra*.

[57] Rouse, Margaret (2023). "What does 256 Bit Encryption mean." Techopedia https://www.techopedia.com/definition/29703/256-bit-encryption.

Algorithm 256-bit), are widely recognized for their robustness and resistance to cryptographic attacks. The 256-bit key length provides an immense number of possible combinations—specifically, 2^256 unique keys—which makes brute-force attacks computationally infeasible with current technology.

The strength of 256-bit security lies in the sheer magnitude of possible key combinations, which renders traditional brute-force attacks impractical. Even with the most powerful supercomputers available today, attempting to exhaustively test all possible keys would require an astronomical amount of time and computational resources. As a result, 256-bit encryption is widely regarded as virtually unbreakable within the foreseeable future, making it an essential component of modern cryptographic systems.

One of the key benefits of 256-bit security is its ability to provide robust protection for sensitive data across various digital environments.[58] Whether encrypting data stored on a computer, transmitting information over the internet, or securing communication channels between devices, 256-bit encryption ensures that data remains confidential and tamper-proof. This level of security is particularly crucial in industries where data privacy and confidentiality are paramount, such as finance, healthcare, and government.

Moreover, the adoption of 256-bit security has become increasingly prevalent as organizations recognize the importance of safeguarding against cyber threats and data breaches.[59] Regulatory requirements and industry standards often mandate the use of strong encryption techniques, including 256-bit encryption, to ensure compliance with

---

[58] *Id.*
[59] *Id.*

data protection regulations and mitigate the risk of security incidents.[60]

In summary, 256-bit security represents a cornerstone of modern cryptography, offering a high level of protection against unauthorized access and data compromise. By leveraging cryptographic algorithms with 256-bit keys, organizations can establish robust security measures to safeguard sensitive information and maintain the integrity of their digital assets.[61] As cyber threats continue to evolve, the adoption of 256-bit encryption remains essential for maintaining confidentiality, integrity, and authenticity in an increasingly interconnected world.

Password Policies

Ensuring a strong password policy is a fundamental aspect of cybersecurity for any organization, as passwords serve as the first line of defense against unauthorized access to sensitive data and systems. A comprehensive password policy encompasses several key factors aimed at promoting robust security practices and mitigating the risk of password-related vulnerabilities.

- **Complexity Requirements**: A strong password policy should mandate the use of complex passwords that are difficult for attackers to guess or crack.[62] This typically involves requiring passwords to contain a combination of uppercase and lowercase letters, numbers, and special characters.

---

[60] Staff (2024). "256-Bit Encryption" Hashed Out at https://www.thesslstore.com/blog/what-is-256-bit-encryption/.
[61] Rouse, *supra.*
[62] Staff (2024). "Best Password Practices." UC Santa Barbra Information Technology https://www.it.ucsb.edu/general-security-resources/password-best-practices.

By enforcing complexity requirements, organizations can significantly increase the strength of passwords and reduce the likelihood of successful brute-force or dictionary attacks.[63]

- **Length Recommendations**: In addition to complexity, password policies should specify minimum length requirements to ensure that passwords offer sufficient entropy and resistance to cracking techniques.[64] Longer passwords are generally more secure than shorter ones, as they increase the number of possible combinations and make it harder for attackers to guess or brute-force the correct password.[65] Recommending or enforcing a minimum password length, such as 12 or 16 characters, helps bolster security against password-based attacks.
- **Multi-Factor Authentication (MFA):** Multi-factor authentication adds an extra layer of security by requiring users to provide additional verification beyond just a password.[66] Common factors include something the user knows (password), something they have (e.g., a smartphone or hardware token), or something they are (biometric data). By implementing MFA alongside traditional password authentication, organizations can significantly enhance security and mitigate the risk of unauthorized access, even if passwords are compromised.[67]

---

[63] Staff (2024). "Use Strong Passwords." Cybersecurity & Infrastructure Security Agency at https://www.cisa.gov/secure-our-world/use-strong-passwords.
[64] UC Santa Barbra, *supra.*
[65] *Id.*
[66] Schrader, Dirk (2023). "The Ultimate Guide to Password Best Practices: Guarding Your Digital Identity." Netwrix at https://blog.netwrix.com/2023/11/15/password-best-practices/.

- **Account Lockout Policies:** Account lockout policies can help defend against brute-force attacks by automatically locking user accounts after a specified number of failed login attempts within a certain time frame.[68] By temporarily locking out accounts, organizations can thwart attackers attempting to guess passwords through automated means while alerting administrators to potential security threats. However, organizations should balance security with user experience and implement measures to prevent denial-of-service attacks resulting from excessive account lockouts.[69]
- **Education and Awareness:** Beyond technical measures, promoting password hygiene through user education and awareness programs is essential for fostering a culture of security within an organization.[70] Employees should receive training on the importance of strong passwords, password best practices, and the risks associated with password reuse, sharing, or storing passwords insecurely.[71] By empowering users to understand and adhere to password policies, organizations can strengthen their overall security posture and reduce the likelihood of successful attacks stemming from weak or compromised passwords.

A robust password policy is essential for safeguarding against password-related vulnerabilities and enhancing overall cybersecurity posture. By incorporating factors such as complexity requirements, length recommendations, regular password changes, multi-factor authentication,

---

[67] *Id.*
[68] CISA, *supra.*
[69] UC Santa Barbra, *supra.*
[70] Schrader, *supra.*
[71] *Id.*

account lockout policies, and user education, organizations can establish effective defenses against unauthorized access and mitigate the risk of password-based attacks.[72] However, password policies should be regularly reviewed and updated to adapt to evolving security threats and industry best practices.

Changing Passwords for Security

Implementing a password changing policy is a critical component of cybersecurity strategy to mitigate the risk of unauthorized access and strengthen overall defenses. While regular password changes can enhance security by reducing the window of opportunity for attackers to exploit compromised credentials, organizations must adopt best practices to ensure the effectiveness and practicality of their password changing policies.

- **Frequency and Interval:** Determining the frequency and interval of password changes is a key consideration in crafting an effective policy.[73] Organizations should strike a balance between security requirements and user convenience, avoiding overly frequent changes that may lead to password fatigue and decreased productivity. A common practice is to set password expiration intervals, such as every 90 or 180 days, to prompt users to update their passwords regularly without causing undue disruption to their workflow.[74]
- **Notification and Reminders:** Providing users with advance notification and reminders before their passwords expire helps promote awareness and

[72] CISA, *supra.*
[73] Rouse, *supra.*
[74] *Id.*

compliance with the password changing policy. Organizations can implement automated email notifications or in-app alerts to inform users of upcoming password expiration dates and prompt them to change their passwords proactively.[75] Clear and concise instructions on how to reset passwords should accompany notifications to facilitate the process for users and minimize confusion.

- **Password Complexity and Strength:** Encouraging or enforcing strong password complexity and strength requirements is essential for ensuring the effectiveness of password changes. Passwords should be sufficiently complex and resistant to brute-force or dictionary attacks, incorporating a combination of uppercase and lowercase letters, numbers, and special characters.[76] By mandating strong passwords during password changes, organizations can bolster security and reduce the risk of password-related vulnerabilities.
- **User Education and Support:** Educating users about the rationale behind password changing policies and providing support resources can foster understanding and compliance.[77] Organizations should offer training sessions or informational materials that highlight the importance of regular password changes in enhancing security posture and protecting sensitive data. Additionally, support channels should be readily available to assist users with any issues or questions they encounter during the password change process, ensuring a smooth and seamless experience.

---

[75] Hashed Out, *supra*.
[76] CISA, *supra*.
[77] *Id.*

Implementing a password changing policy requires careful consideration of factors such as frequency, notification, complexity, and user support to strike a balance between security requirements and user experience. By adopting best practices for password changes, organizations can mitigate the risk of unauthorized access and strengthen their overall cybersecurity posture. However, it is essential to regularly review and update password changing policies in response to evolving threats and industry standards to maintain their effectiveness over time.

Conclusion

In conclusion, effective password policies are paramount in safeguarding sensitive information and mitigating cybersecurity risks in today's digital landscape. The evolution of password security, from rudimentary practices to sophisticated measures, underscores the ongoing battle against cyber threats and the need for continuous adaptation. The concept of 256-bit security epitomizes the robust encryption standards that modern organizations employ to protect their data from unauthorized access and exploitation.[78] By implementing stringent password policies that emphasize complexity, strength, and regular changes, companies can bolster their defenses and thwart malicious actors' attempts to compromise their systems.

Furthermore, password changing policies must strike a balance between security requirements and user convenience to promote adherence and minimize friction. Clear communication, user education, and adequate support mechanisms are essential components of successful password policies, fostering awareness and compliance among employees. As organizations navigate the

---

[78] Hashed Out, *supra.*

complexities of cybersecurity, they must prioritize password hygiene as a fundamental aspect of their defense strategy. With a robust password policy framework in place, businesses can fortify their digital fortresses and mitigate the ever-present threats posed by cyber adversaries.

Chapter 3: Data Storage and Devices

"In the next three years, the value of data will increase, making it even more valuable than it is today. The more efficiently you store your data, the more benefits your business will see."

Thomas Harrer[79]

This chapter explores the critical aspects of data storage and security, highlighting the importance of safeguarding sensitive information stored on various devices. The chapter begins by examining the diverse options available for data storage devices, ranging from traditional hard drives to modern cloud storage solutions. Understanding the strengths and vulnerabilities of each storage option is essential for organizations to make informed decisions regarding data protection and accessibility.

One key aspect discussed in this chapter is the importance of locking data storage devices, such as USB dongles or external hard drives, to prevent unauthorized access or data theft.[80] Implementing robust encryption and access controls ensures that sensitive data remains secure, even if the physical device falls into the wrong hands. By adopting encryption standards and encryption protocols, organizations can bolster the confidentiality and integrity of their data storage solutions. Additionally, we consider the enduring relevance of traditional storage mediums like floppy disks and CDs, highlighting their utility in certain

[79] (2024). "Data Storage Quotes, Thomas Harrer." Good Reads at https://www.goodreads.com/quotes/tag/data-storage.
[80] Ashtari, Hossein (2022). "What is Hardware Security? Definition, Threats, and Best Practices." Spiceworks at https://www.spiceworks.com/it-security/vulnerability-management/articles/what-is-hardware-security/.

scenarios and the need for proper handling and storage to prevent data degradation or loss.[81] While these mediums may seem outdated in the age of cloud storage, they still serve as viable backup options for critical data and offer an additional layer of redundancy in data management strategies.

We emphasizes the importance of implementing effective backup systems to mitigate the risk of data loss due to hardware failures, cyberattacks, or other unforeseen circumstances. By adopting robust backup practices, such as regular backups, offsite storage, and versioning, organizations can ensure data resilience and continuity in the face of potential threats or disruptions. This chapter emphasizes the significance of data storage security and the need for organizations to adopt comprehensive strategies to protect their valuable information assets. By leveraging a combination of secure storage devices, encryption technologies, and robust backup systems, organizations can safeguard against data breaches, ensure business continuity, and maintain the integrity and confidentiality of their data in an increasingly digitalized world.

Data Storage Device Options

Several data storage device options are available on the market, each offering varying levels of security and functionality. Some of the best and safest options include:

1. **Solid State Drives (SSDs):** SSDs have gained popularity due to their fast read and write speeds,

[81] Ahmadvand, Mohsen (2019) "A Taxonomy of Software Integrity Protection Techniques." Hardware Security at https://www.sciencedirect.com/topics/computer-science/hardware-security.

reliability, and durability. They use flash memory to store data, making them less susceptible to mechanical failures compared to traditional hard disk drives (HDDs).[82] SSDs also support encryption features, enhancing data security.

2. **Hardware-Encrypted USB Drives:** These portable storage devices feature built-in encryption capabilities, ensuring that data remains protected even if the drive is lost or stolen.[83] Users can set up password protection and encryption keys to prevent unauthorized access to the stored data.
3. **Network-Attached Storage (NAS):** NAS devices provide centralized storage solutions for homes and businesses, allowing multiple users to access and share data over a network. NAS systems often come with RAID (Redundant Array of Independent Disks) configurations for data redundancy and fault tolerance, enhancing data protection.[84]
4. **Cloud Storage Services:** Cloud storage offers convenient and scalable storage solutions, allowing users to store data remotely on servers managed by third-party providers.[85] Leading cloud storage providers implement robust security measures, such as encryption, access controls, and data replication, to safeguard user data against unauthorized access and data loss.
5. **External Hard Disk Drives (HDDs):** Despite the growing popularity of SSDs, external HDDs remain a cost-effective option for storing large volumes of data.[86] While HDDs are more prone to physical

---

[82] *Id.*
[83] *Id.*
[84] Staff (2023). "The significance of Computer Hardware Engineering in Cybersecurity." Utilities One at https://utilitiesone.com/the-significance-of-computer-hardware-engineering-in-cybersecurity.
[85] Ashtari, *supra.*

damage and mechanical failures compared to SSDs, they still offer reliable storage solutions for backup and archival purposes when handled with care.[87]

6. **Write-Once Read-Many (WORM) Drives:** WORM drives are specialized storage devices designed for long-term data retention and regulatory compliance.[88] They allow data to be written once and then read multiple times, preventing tampering or alteration of stored information.
7. **Tape Storage:** Tape storage remains a viable option for archival and backup purposes, offering high capacity and low cost per gigabyte.[89] Modern tape storage solutions feature encryption capabilities and robust error correction mechanisms to ensure data integrity and security.[90]

Ultimately, the best data storage device option depends on factors such as data security requirements, storage capacity needs, performance considerations, and budget constraints. Organizations should carefully evaluate their specific use cases and compliance requirements when selecting the most suitable storage solution for their needs.

Lock Your Dongles (Or Hard Drives)

Locking dongles (thumb drives) or hard drives is paramount for maintaining the security and integrity of sensitive data in both personal and professional settings. These storage devices are often used to store and transfer

---

[86] Staff (2024). "What is Hardware and Software Security?" Wheelhouse at https://www.wheelhouse.com/resources/what-is-hardware-and-software-security-a11018.

[87] *Id*,

[88] Ahmadvand, *supra*.

[89] *Id*.

[90] Wheelhouse, *supra*.

critical information, ranging from confidential business documents to personal financial records.[91] Failure to secure these devices can expose the stored data to various security risks, including unauthorized access, data theft, and malware infections.

One of the primary reasons for locking dongles or hard drives is to prevent unauthorized access to the stored data in the event of loss or theft. These portable storage devices are highly susceptible to being misplaced, stolen, or accidentally left behind in public spaces. Without proper security measures in place, anyone who finds or steals the device could potentially access its contents, compromising sensitive information and exposing individuals or organizations to data breaches and legal liabilities.

Furthermore, locking dongles or hard drives adds an extra layer of protection against malware and cyber threats. Malicious actors often target removable storage devices as a vector for spreading malware, such as viruses, ransomware, and spyware.[92] By encrypting or password-protecting these devices, users can mitigate the risk of unauthorized parties tampering with the stored data or introducing malicious software, thus safeguarding the confidentiality, integrity, and availability of their information assets.

Good Old Fashion Floppies

Classic mediums such as CDs and floppies have been used for decades as storage solutions for various types of data. Despite the advent of more advanced storage technologies, these legacy mediums continue to serve a purpose in

---

[91] Ashtari, *supra*.
[92] *Id.*

certain scenarios due to their unique advantages.[93] One of the primary benefits of CDs and floppies is their offline nature, which makes them immune to online cyber threats such as hacking and remote data breaches. This offline characteristic makes them ideal for storing sensitive information that requires enhanced security measures.

Moreover, CDs and floppies offer a level of physical durability and longevity that digital storage solutions cannot always guarantee. Unlike hard drives or solid-state drives (SSDs), which are susceptible to mechanical failure or data corruption over time, CDs and floppies are relatively resilient to physical damage and can withstand harsh environmental conditions.[94] This durability makes them suitable for archival purposes, where long-term data retention is essential.

Additionally, CDs and floppies provide a convenient means of data portability and compatibility across different systems and devices. Most modern computers still feature CD/DVD drives or support USB floppy disk drives, allowing users to easily access and transfer data stored on these mediums. This compatibility ensures that data stored on CDs and floppies can be accessed and utilized without the need for specialized hardware or software, making them accessible to a wide range of users.

However, despite their usefulness, it is crucial to recognize the limitations and security risks associated with classic mediums like CDs and floppies. As legacy technologies, they may lack the robust encryption and security features

---

[93] Korolov, Maria (2015). "Significant Virtual Machine Vulnerability has been hiding in Floppy Disk code for 11 years." CSO at https://www.csoonline.com/article/551445/significant-virtual-machine-vulnerability-has-been-hiding-in-floppy-disk-code-for-11-years.html.
[94] *Id.*

found in modern storage solutions, making them vulnerable to unauthorized access or data breaches.[95] Additionally, their relatively small storage capacities compared to newer mediums may not be sufficient for storing large volumes of data or high-definition multimedia content.

Furthermore, when CDs and floppies reach the end of their lifecycle or are no longer needed, proper disposal is essential to prevent potential data breaches or environmental hazards. Unlike digital data that can be securely erased or overwritten, physical storage mediums like CDs and floppies require physical destruction to ensure that the data they contain cannot be recovered or accessed by unauthorized parties. Failure to dispose of these mediums properly could expose sensitive information to malicious actors or result in the inadvertent leakage of confidential data.

While classic mediums like CDs and floppies offer certain advantages in terms of offline security, durability, and compatibility, they must be used judiciously and disposed of properly to mitigate security risks. Organizations and individuals should implement appropriate measures to safeguard data stored on these mediums and ensure compliance with data protection regulations.[96] Additionally, transitioning to modern storage solutions with enhanced security features can help address the evolving threat landscape and protect against emerging cyber threats.

---

95 Staff (2022). "What to do with old floppy disks." Cultrix at https://www.cultrix.co.uk/it-academy/what-to-do-with-old-floppy-discs/.

96 Korolov, *supra.*

Backup Systems

Securing backup systems is paramount for safeguarding critical data and ensuring business continuity in the face of unforeseen events such as hardware failures, natural disasters, or cyberattacks.[97] Backup systems serve as a safety net by creating duplicate copies of essential data, applications, and configurations, allowing organizations to recover swiftly and minimize downtime in the event of data loss or system failures.

First, backup systems provide redundancy, which is essential for mitigating the risk of data loss due to hardware malfunctions or system failures. By regularly backing up data to offsite or cloud-based storage repositories, organizations can ensure that they have multiple copies of their data stored in different locations, reducing the likelihood of complete data loss in the event of a catastrophic failure.

Second, backup systems play a crucial role in protecting against cyber threats such as ransomware, malware, and data breaches.[98] In the event of a cyberattack, where data may be encrypted or compromised, having recent backups enables organizations to restore their systems to a pre-attack state and recover data without capitulating to the demands of cybercriminals. This capability is particularly vital in industries where data integrity and confidentiality are paramount, such as healthcare, finance, and government sectors.

Next, securing backup systems involves implementing robust access controls and encryption mechanisms to

---

[97] Ashtari, *supra*.
[98] *Id.*

prevent unauthorized access or tampering with backup data. By restricting access to backup repositories and employing strong encryption algorithms to protect data both in transit and at rest, organizations can safeguard sensitive information from insider threats and external adversaries seeking to exploit vulnerabilities in backup infrastructure.[99] Finally, backup systems should undergo regular testing and validation to ensure their reliability and effectiveness in the event of a data loss scenario. Conducting routine backup tests, including backup integrity checks, data restoration drills, and disaster recovery simulations, allows organizations to identify and address any deficiencies or vulnerabilities in their backup processes before they affect critical operations.

Securing backup systems is a primary component of an organization's overall cybersecurity strategy, providing resilience, redundancy, and data protection against a wide range of threats and vulnerabilities.[100] By implementing comprehensive backup solutions with robust security controls, organizations can enhance their ability to recover from data loss events, maintain operational continuity, and safeguard their most valuable asset – their data.

Conclusion

In conclusion, chapter 3 reviews the critical aspects of data storage and backup systems, emphasizing the importance of adopting robust security measures to protect sensitive information and ensure business continuity. The chapter highlights various data storage device options, including cloud-based solutions, external hard drives, and solid-state

[99] Ahmadvand, *supra.*
[100] Wheelhouse, *supra.*

drives, each offering distinct advantages in terms of reliability, accessibility, and scalability.

Furthermore, the chapter underscores the significance of securing data storage devices through measures such as encryption, access controls, and regular backups. By implementing stringent security protocols, organizations can safeguard their data against unauthorized access, data breaches, and cyber threats, thereby minimizing the risk of data loss and ensuring compliance with regulatory requirements.

Moreover, the chapter underscores the importance of securely locking dongles and hard drives to prevent unauthorized access and protect sensitive data from theft or tampering. By employing encryption technologies and physical security measures, organizations can mitigate the risk of data breaches and safeguard their intellectual property and confidential information.

Overall, this chapter underscores the critical role of data storage and backup systems in maintaining data integrity, confidentiality, and availability. By adopting a proactive approach to data security and implementing robust security measures across all aspects of data storage and backup infrastructure, organizations can enhance their resilience to cyber threats, mitigate the risk of data loss, and safeguard their most valuable asset – their data.

Chapter 4: Access to Sensitive Locations

"It takes a lot of courage to release the familiar and seemingly security, to embrace the new. But there is no real security in what is no longer meaningful. There is more security in the adventurous and exciting, for in movement there is life, and in change there is power."

Alan Cohen[101]

In this chapter, we analyzes the critical aspects of access control and perimeter security within organizational settings, emphasizing the importance of robust measures to safeguard physical locations and assets from unauthorized access or breaches. The chapter begins by addressing the security conundrum inherent in scenarios where individuals lacking proper clearance may inadvertently compromise security protocols or where surveillance of high-security areas may be inadequate. It underscores the necessity of implementing stringent access control mechanisms and surveillance systems to mitigate such risks effectively.

A key focus of the chapter lies in examining the phenomenon of tailgating, wherein unauthorized individuals gain entry to secure areas by closely following authorized personnel. Through comprehensive awareness programs and the deployment of access control technologies like biometric authentication or turnstile gates, organizations can mitigate the risk of tailgating incidents and reinforce perimeter security. Additionally, the concept of creating rings of security is explored, wherein access levels are hierarchically structured based on the sensitivity of information or assets, with stricter protocols enforced closer to critical areas.

---

[101] (2023). "Security Quotes, Alan Cohen." Brainy Quote at https://www.brainyquote.com/topics/security-quotes.

Furthermore, the chapter underscores the distinction between clearance and access, advocating for a need-to-know approach wherein access privileges are granted based on legitimate business requirements rather than broad clearances.[102] By adopting this principle, organizations can minimize the likelihood of insider threats and unauthorized disclosures while enhancing accountability and transparency in access management practices. Additionally, the chapter emphasizes the significance of regular audits, monitoring, and training initiatives to ensure compliance with access control policies and to identify and rectify potential vulnerabilities proactively.

Here we underscore the critical role of access control in bolstering organizational security posture and mitigating risks associated with unauthorized access or intrusions. By implementing robust access control mechanisms, leveraging advanced surveillance technologies, and promoting a culture of security awareness, organizations can fortify their defenses, safeguard physical assets, and uphold the confidentiality and integrity of sensitive information. Adopting a multi-layered approach to access management, coupled with stringent enforcement of clearance and access protocols, is essential for maintaining a secure environment conducive to business operations and resilience against evolving threats.

The Security Conundrum (People without clearance cannot watch those who do have clearance).

In the realm of security, the conundrum often arises from the balance between access and oversight, particularly in

---

[102] Koek, Bart (2023). "Best Practices for effectively securing sensitive data." Security at https://www.securitymagazine.com/articles/99931-best-practices-for-effectively-securing-sensitive-data.

environments where individuals without clearance may inadvertently gain visibility into restricted areas, or worse, where no supervision is present to monitor access.[103] This presents a multifaceted challenge for organizations striving to maintain stringent security protocols. On one hand, having individuals lacking clearance observe those with authorized access introduces the risk of information leakage or compromise.[104] Conversely, the absence of supervision altogether leaves the door open for unauthorized entry or activities to go unnoticed, potentially leading to security breaches or data breaches.

Addressing the security conundrum necessitates a comprehensive approach that encompasses both technological solutions and stringent procedural protocols. One key consideration is implementing robust access control measures, such as biometric authentication or access card systems, to restrict entry to only those with proper clearance.[105] Additionally, employing surveillance systems and security personnel can help mitigate the risk of unauthorized observation or access by providing constant monitoring and response capabilities.

However, it's essential to recognize that certain support staff, such as custodians or maintenance personnel, may require periodic access to high-security areas to perform their duties. In such cases, strict protocols must be established to ensure that access is granted only when

---

[103] Li, Amanda (2023). "Top 8 Pros and Cons of Surveillance Cameras in Public Places." Redlink at https://reolink.com/blog/pros-cons-of-surveillance-cameras-in-public-places/.
[104] *Id.*
[105] Bergengruen, Vera (2023). "Why the U.S. Security-Clearance Process has a Digital Blind Spot." Time at https://time.com/6276737/jack-teixeira-security-clearance-online-behavior/.

necessary and under supervision.[106] This may involve scheduling cleaning or maintenance activities during off-peak hours or implementing escort policies to accompany support staff while they are within restricted areas.

Furthermore, organizations must prioritize training and awareness initiatives to educate all personnel about the importance of maintaining security protocols and the potential consequences of non-compliance. By fostering a culture of security consciousness throughout the organization, employees are more likely to recognize and report any suspicious behavior or security lapses, thereby enhancing overall vigilance and risk mitigation efforts.[107] In essence, addressing the security conundrum requires a proactive and multifaceted approach that combines technological solutions, procedural safeguards, and employee awareness to minimize the risk of unauthorized access or observation while ensuring that essential support functions can still be performed safely and securely.

Tailgating at Work

Tailgating, in the area of security, refers to the unauthorized practice of an individual gaining access to a restricted area by closely following an authorized person through a secure entry point, such as a door or gate.[108] This clandestine tactic capitalizes on the natural inclination of individuals to hold the door open for others behind them, exploiting human behavior to bypass established access controls.[109] Despite its seemingly innocuous nature,

---

[106] Li, *supra*.
[107] *Id.*
[108] Awati, Rahul (2024). "Tailgating (Piggybacking)." TechTarget at https://www.techtarget.com/whatis/definition/tailgating-piggybacking.
[109] Staff (2022). "What are tailgating attacks and how to protect yourself from them." McAfee at

tailgating poses a significant security risk, as it can grant unauthorized individuals entry into sensitive areas where they may perpetrate malicious activities or gain access to confidential information.

To prevent tailgating and uphold the integrity of access control measures, companies must implement a combination of technological solutions and procedural protocols.[110] One effective approach is the deployment of access control systems that incorporate features such as biometric authentication, access cards, or proximity sensors.[111] These systems are designed to ensure that only authorized individuals can gain entry, thereby thwarting attempts at tailgating by unauthorized individuals attempting to piggyback on the access of others.

Additionally, companies should establish clear policies and procedures regarding access to secure areas, emphasizing the importance of vigilance and compliance with security protocols among all personnel. This may include conducting regular training sessions to educate employees about the risks associated with tailgating and the critical role they play in maintaining a secure environment.[112] Furthermore, organizations should encourage a culture of accountability, empowering employees to challenge individuals they do not recognize or who lack proper identification before granting them access.

Moreover, physical security measures such as turnstiles, mantraps, doors in sets of two, or security guards stationed

---

https://www.mcafee.com/blogs/internet-security/what-are-tailgating-attacks/.

[110] Awati, *supra.*

[111] *Id.*

[112] Staff (2023). "What is a Tailgating attack?" Fortra at https://terranovasecurity.com/blog/tailgating-attack/.

at entry points can serve as deterrents to unauthorized access attempts, effectively preventing tailgating incidents.[113] These measures create additional barriers that require individuals to authenticate their identity or obtain proper authorization before gaining entry, reducing the likelihood of unauthorized access.[114] By adopting a multi-layered approach to security that combines technological solutions, procedural safeguards, and physical barriers, companies can effectively mitigate the risk of tailgating and uphold the confidentiality and integrity of their sensitive assets.[115]

Rings of Security

Implementing a hierarchical system of access control, often referred to as "rings of access," is a strategic approach to enhancing security within a company's premises.[116] This concept involves dividing the physical space into concentric rings, with each ring representing a different level of sensitivity or criticality. The outermost ring typically encompasses public or low-security areas, while the inner rings contain increasingly sensitive zones, such as data centers, executive offices, or research laboratories.[117] By assigning varying levels of access privileges to individuals based on their role and clearance level, companies can enforce stricter security protocols in areas with higher sensitivity, thereby reducing the risk of unauthorized access or security breaches.

---

[113] McAfee, *supra.*
[114] Staff (2024). "What is Tailgating?" AwareGo at https://awarego.com/tailgating/.
[115] *Id.*
[116] Staff (2023A), "Securing Geolocation Data Best Practices and Strategies." Utilities One at https://utilitiesone.com/securing-geolocation-data-best-practices-and-strategies.
[117]

In practice, the implementation of rings of access involves establishing clear boundaries between different zones within the facility and deploying appropriate access control measures at each entry point. For example, external doors leading into the building may be equipped with robust security features such as access card readers, biometric scanners, or surveillance cameras to prevent unauthorized entry from the outside.[118] As individuals move inward towards more sensitive areas, additional layers of security measures are deployed, including authentication mechanisms such as PIN codes, security guards, or mantraps.

The key principle behind rings of access is the concept of "defense in depth," which advocates for the layering of multiple security controls to create overlapping layers of protection.[119] By implementing this approach, companies can establish a series of checkpoints or barriers that must be successfully navigated to access increasingly sensitive areas.[120] For instance, individuals may be required to present multiple forms of identification, undergo biometric authentication, or obtain explicit authorization from security personnel before advancing to the next ring.

Companies can leverage technology to enforce access restrictions dynamically based on situational factors or time-sensitive requirements.[121] Access control systems can be programmed to adjust access privileges in real-time, granting temporary access to specific individuals or groups based on predefined criteria such as time of day, job function, or security clearance level. This dynamic

---

118 Utilities One (2023A), *supra.*
119 Staff (2023A). "Data Security Best Practices." Netwrix at https://www.netwrix.com/data-security-best-practices.html.
120 *Id.*
121 Utilities One (2023A), *supra.*

approach ensures that access rights are tailored to the specific needs of employees or visitors while maintaining the integrity of security protocols.

Integrating surveillance and monitoring capabilities into the rings of access can provide valuable insights into employee behavior and identify potential security threats in real-time.[122] Video surveillance cameras strategically positioned at entry points and throughout the facility enable security personnel to monitor and record access attempts, detect anomalies, and respond promptly to security incidents. Additionally, access logs generated by the access control system can be analyzed to identify patterns or trends indicative of unauthorized access attempts or security breaches.

Overall, the implementation of rings of access represents a proactive strategy for enhancing security within an organization by creating concentric layers of protection that align with the sensitivity of the assets or areas being safeguarded. By establishing clear boundaries, deploying appropriate access control measures, and leveraging technology for dynamic access management and surveillance, companies can mitigate the risk of unauthorized access, safeguard critical assets, and maintain a secure environment for employees, visitors, and stakeholders.

Clearance and Access (Need to Know)

Clearance and access are two distinct concepts in the realm of security management, each playing a crucial role in safeguarding sensitive information and assets within an

[122] Mohan, Sanjeev (2023). "Best Practices for Securing Sensitive Data." Medium at https://sanjmo.medium.com/best-practices-for-securing-sensitive-data-f52a1eda8593.

organization. Clearance refers to the formal authorization granted to an individual, typically following a thorough background check or vetting process, to access classified or restricted information based on their level of trustworthiness, reliability, and suitability for handling sensitive data. Clearance levels are often categorized hierarchically, such as confidential, secret, or top-secret, depending on the sensitivity of the information involved.

On the other hand, access pertains to the physical or digital means by which an individual is granted entry to specific areas, systems, or resources within the organization's premises or network infrastructure.[123] Unlike clearance, which is based on an individual's background, qualifications, and role within the organization, access privileges are determined by the specific job responsibilities, functional requirements, or operational needs of the individual in question.[124] In essence, access defines what a person is allowed to do or interact with, while clearance determines the level of sensitivity or confidentiality of the information they are permitted to access.

To maintain a robust security posture and mitigate the risk of unauthorized disclosure or misuse of sensitive information, companies should adopt a "need-to-know" principle when granting access privileges to employees. Under this principle, access rights are strictly limited to the minimum level necessary for individuals to perform their job functions effectively.[125] For instance, employees working on a particular project may only be granted access to relevant documents or resources directly related to their

---

[123] *Id.*
[124] Utilities One (2023A), *supra.*
[125] Mohan, *supra.*

assigned tasks, while access to higher-level information or confidential data is restricted to a select group of authorized personnel with a demonstrated need for such information.

By implementing a need-to-know access control framework, companies can minimize the potential for data breaches, insider threats, or inadvertent exposure of sensitive information to unauthorized individuals.[126] This approach ensures that access privileges are aligned with job responsibilities and operational requirements, reducing the likelihood of data leakage or security incidents resulting from over-privileged access.[127] Additionally, regular audits and reviews of access permissions are essential to identify and address any discrepancies, unauthorized accesses, or security vulnerabilities that may compromise the integrity of the access control system.

Companies should prioritize security awareness training and education to promote a culture of responsibility, accountability, and confidentiality among employees.[128] By raising awareness about the importance of safeguarding sensitive information, adhering to access control policies, and reporting any suspicious activities or security concerns, employees can become proactive participants in the organization's security efforts.[129] Through a combination of robust access control mechanisms, stringent clearance procedures, and ongoing security awareness initiatives, companies can establish a comprehensive security framework that effectively protects critical assets, preserves

---

[126] *Id.*.
[127] *Id.*
[128] Slavin, Brad (2022). "9 Best Practices to Manage Sensitive Data Carefully." DuoCircle at https://www.duocircle.com/email-security/9-best-practices-to-manage-sensitive-data-carefully.
[129] *Id.*

confidentiality, and mitigates the risk of unauthorized access or data breaches.

Conclusion

In conclusion, securing access to physical locations within an organization is paramount to maintaining a robust security posture and safeguarding sensitive assets against unauthorized access or intrusions. Throughout this chapter, we have explored various strategies and best practices aimed at mitigating security risks associated with access control and perimeter protection. From addressing the security conundrum of individuals without clearance having access to sensitive areas to implementing measures to prevent tailgating incidents, organizations must adopt a multi-layered approach to access management.

By establishing rings of security that delineate access levels based on the sensitivity of information and the criticality of resources, companies can create effective barriers against unauthorized entry and mitigate the risk of insider threats. Moreover, the implementation of clearance-based access control mechanisms, coupled with a need-to-know principle, ensures that access privileges are granted only to individuals with a legitimate business need for accessing specific resources or information. This approach not only reduces the attack surface but also enhances accountability and transparency in access management practices.

Furthermore, organizations should prioritize regular audits, monitoring, and training initiatives to ensure compliance with access control policies, identify potential vulnerabilities, and reinforce security awareness among employees. By continuously refining access control procedures, updating security protocols, and leveraging technology solutions such as biometric authentication or access control systems, companies can adapt to evolving security threats and maintain a proactive stance in safeguarding their physical assets and intellectual property. In essence, effective access control measures serve as a cornerstone of comprehensive security strategies, enabling organizations to fortify their defenses, mitigate risks, and uphold the confidentiality, integrity, and availability of critical resources.

Conclusion Part I

Part I of the book has provided a comprehensive exploration of foundational principles and strategies essential for establishing a secure operational environment within organizations. From securing individual workstations to implementing robust access control measures, each chapter has delved into critical aspects of cybersecurity and physical security, underscoring the importance of proactive measures to mitigate risks and safeguard assets. As we conclude this section, it is evident that the landscape of security in the Web3 era demands a multifaceted approach that addresses both digital and physical vulnerabilities with equal rigor.

The chapters in Part I have emphasized the notion that an individual's computer is their fortress, highlighting the importance of stringent login procedures, access controls, and monitoring systems to safeguard against unauthorized access and potential threats. By instilling a culture of responsibility and accountability among employees regarding their workstation security, organizations can fortify their defenses and mitigate the risk of data breaches or intrusions.

Furthermore, the discussion on password policies has underscored the critical role of robust authentication mechanisms in protecting sensitive information and assets from unauthorized access. Through the adoption of advanced encryption techniques and stringent password policies, organizations can enhance the security posture of their digital infrastructure and thwart cyber threats effectively.

Moreover, the exploration of data storage device options and backup systems has shed light on the importance of implementing secure storage solutions and contingency plans to prevent data loss or theft. By leveraging encryption technologies, access controls, and regular backups, organizations can ensure the integrity and availability of their data assets, even in the face of unforeseen incidents or cyberattacks.

The chapters on access control and perimeter security have emphasized the significance of implementing comprehensive access management protocols and surveillance systems to safeguard physical locations and assets from unauthorized access or breaches. By adopting a multi-layered approach to access control, organizations can mitigate the risk of insider threats, tailgating incidents, and unauthorized disclosures, thereby bolstering overall security posture.

In conclusion, Part I has provided a holistic framework for addressing the myriad challenges and complexities associated with security in the Web3 era. By integrating the principles and strategies outlined in these chapters into their security practices, organizations can establish a robust security foundation, safeguard their digital and physical assets, and navigate the evolving threat landscape with confidence and resilience. As we proceed into subsequent sections, we will continue to explore advanced concepts and emerging technologies aimed at further enhancing security in the digital age.

Moreover, the exploration of data storage device options and backup systems has shed light on the importance of implementing secure storage solutions and contingency plans to prevent data loss or theft. By leveraging encryption technologies, access controls, and regular backups, organizations can ensure the integrity and availability of their data assets, even in the face of unforeseen incidents or cyberattacks.

The chapters on access control and perimeter security have emphasized the significance of implementing comprehensive access management protocols and surveillance systems to safeguard physical locations and assets from unauthorized access or breaches. By adopting a multi-layered approach to access control, organizations can mitigate the risk of insider threats, tailgating incidents, and unauthorized disclosures, thereby bolstering overall security posture.

In conclusion, Part I has provided a holistic framework for addressing the myriad challenges and complexities associated with security in the Web3 era. By integrating the principles and strategies outlined in these chapters into their security practices, organizations can establish a robust security foundation, safeguard their digital and physical assets, and navigate the evolving threat landscape with confidence and resilience. As we proceed into subsequent sections, we will continue to explore advanced concepts and emerging technologies aimed at further enhancing security in the digital age.

Part II: Account Security

Introduction Part II

Part II of this book examines the domain of account security within the dynamic landscape of web3 technologies. In this section, we embark on a comprehensive exploration of the multifaceted challenges and proactive strategies essential for safeguarding digital assets, protecting sensitive data, and mitigating emerging threats in the digital realm. From password management and encryption protocols to incident response planning and insider threat mitigation, each chapter offers invaluable insights and practical recommendations for fortifying the security posture of web3 companies.

As the digital landscape continues to evolve at a rapid pace, the significance of robust account security measures cannot be overstated. In an era characterized by decentralized technologies, blockchain innovations, and interconnected digital ecosystems, organizations must remain vigilant against a myriad of cyber threats, ranging from sophisticated phishing attacks to insider breaches. This part of this book serves as a comprehensive guide for navigating the complex terrain of web3 security, equipping readers with the knowledge and tools necessary to confront these challenges head-on.

The chapters in this section are crafted to address key aspects of account security within the context of web3 technologies. From understanding the nuances of password policies and encryption algorithms to implementing effective incident response strategies and mitigating insider threats, each topic is explored in depth to provide readers with a holistic understanding of the intricacies involved in safeguarding digital assets and mitigating cyber risks.

Furthermore, Part II underscores the importance of proactive risk management and continuous improvement in the realm of web3 security. With the proliferation of advanced cyber threats and the increasing sophistication of malicious actors, organizations must adopt a proactive stance towards security, leveraging cutting-edge technologies, robust protocols, and best practices to stay ahead of evolving threats and vulnerabilities.

Moreover, the chapters in this section emphasize the critical role of collaboration and knowledge-sharing in bolstering web3 security. By fostering a culture of collaboration among stakeholders, including cybersecurity professionals, IT teams, and business leaders, organizations can harness collective expertise and resources to strengthen their defenses, detect and respond to threats more effectively, and uphold the integrity and trust of their digital ecosystems.

In summary, Part II serves as an indispensable resource for organizations seeking to enhance their account security posture in the web3 era. Through a comprehensive exploration of password management, encryption protocols, incident response planning, insider threat mitigation, and other key aspects of web3 security, this section empowers readers with the knowledge and tools needed to navigate the complex landscape of digital risk and safeguard their digital assets in an increasingly interconnected world.

Chapter 5: Password Storage

"The whole notion of passwords is based on an oxymoron. The idea is to have a random string that is easy to remember. Unfortunately, if it's easy to remember, it's something nonrandom like 'Susan.' And if its random, like 'r7U2*Qnp,' then it's not easy to remember."

Bruce Scheier[130]

Chapter 5 examines the intricate landscape of password storage issues within Web3 companies, shedding light on the multifaceted challenges and vulnerabilities inherent in authentication practices. As organizations navigate the complexities of modern cybersecurity, they encounter hurdles related to the memorability of complex passwords, the protection of public and private keys, the security of storage software, and the risks associated with dual-factor authentication on the same device.

One of the primary hurdles discussed is the difficulty users face in remembering longer and more complex passwords.[131] As security requirements evolve, the demand for stronger passwords increases, posing a cognitive burden on users and potentially leading to the adoption of weaker or more easily guessable credentials, undermining overall security posture.

Another critical aspect examined is the importance of safeguarding public and private keys, which serve as the

---

[130] (2024). "Security Quotes, Bruce Schneier." AZ Quotes at https://www.azquotes.com/quotes/topics/passwords.html.

[131] Staff (2024). "Cybersecurity 101: Why Choosing a Secure Password is So Important." Walden University at https://www.waldenu.edu/programs/information-technology/resource/cybersecurity-101-why-choosing-a-secure-password-in-so-important.

linchpin of cryptographic security in Web3 environments.[132] While public keys are meant for widespread dissemination, the exposure of private keys can have catastrophic consequences, leading to unauthorized access, data breaches, and financial losses. Hence, the imperative to keep private keys confidential is underscored.

Furthermore, the chapter underscores the risks associated with storage software used for managing passwords and keys.[133] Despite offering convenience and accessibility, such software can become a single point of failure if compromised, leading to widespread data exposure and exploitation. Organizations must adopt stringent measures to secure their storage solutions and mitigate the inherent risks.

Additionally, the dangers of using the same device for both data access and two-factor authentication are explored, highlighting the inherent vulnerabilities in such practices. From malware attacks to physical theft, the reliance on a single device for multiple functions increases the attack surface and heightens the risk of unauthorized access, necessitating a more robust approach to security.[134]

In essence, this chapter emphasizes the critical need for Web3 companies to address password storage issues with

---

[132] Staff (2024). "What are Public and Private Keys." AppveiwX at https://www.appviewx.com/education-center/what-are-public-and-private-keys/.

[133] (N/A) (2024). "The Hidden Dangers of Storing Passwords in Your Browser." MainStreet Solutions at https://www.mainstreetitsolutions.com/the-hidden-dangers-of-storing-passwords-in-your-browser/.

[134] Devi, Navanita (2023). "How Secure is Two-Factor Authentication (2FA)?" LoginRadius at https://www.loginradius.com/blog/identity/how-secure-2fa/.

diligence and foresight, implementing robust security measures, and adopting best practices to safeguard sensitive information, protect cryptographic assets, and mitigate the risks associated with authentication in an increasingly digital and interconnected landscape.

Difficulty Remembering Complex Passwords

Longer, more complex passwords have become a necessity in the age of web3 security, where the threat landscape is ever-evolving and cyberattacks are increasingly sophisticated.[135] However, despite the critical importance of robust password practices, many individuals struggle with remembering these complex credentials. One of the primary reasons for this difficulty lies in the human brain's limited capacity for retaining and recalling information, especially when it comes to arbitrary strings of characters that lack personal significance or context.[136]

Moreover, the proliferation of online accounts across various platforms and services exacerbates the challenge of password management. With the average internet user juggling multiple accounts, each requiring a unique and sufficiently complex password, the cognitive load associated with memorization becomes overwhelming.[137] As a result, individuals often resort to using simpler, more memorable passwords or reusing the same password across multiple accounts, unwittingly compromising their security posture.

---

135 Schneier, *supra*.
136 Cherrie, Chrysta (2023). "The Life and Death of Passwords: The Problem with Passwords." Cisco Duo at https://duo.com/blog/life-death-passwords-problem-with-passwords.
137 *Id.*

The human tendency to favor convenience over security further compounds the issue of password memorability. In an era characterized by instant gratification and user-centric design, individuals are naturally inclined to prioritize ease of access and frictionless interactions.[138] This inclination often leads them to opt for passwords that are easy to recall or use familiar patterns and phrases, disregarding the importance of strong, randomly generated passwords in thwarting brute-force attacks and unauthorized access attempts.[139]

Additionally, the inherent complexity of modern password requirements, which often mandate a combination of uppercase and lowercase letters, numbers, and special characters, presents a formidable cognitive challenge for many users.[140] As password complexity increases, so does the mental effort required to create, remember, and accurately input these intricate strings of characters. This cognitive overhead can be particularly daunting for individuals with cognitive impairments, learning disabilities, or age-related memory decline, further exacerbating the accessibility and usability issues surrounding password management.

Furthermore, the relentless pace of technological advancement and the constant evolution of cybersecurity threats contribute to a sense of password fatigue among users.[141] With each data breach or security incident making headlines, individuals are bombarded with warnings about the importance of strong passwords and the risks of

---

[138] List, Jenny (2023). "The Problem with Passwords." Hackaday at https://hackaday.com/2023/01/03/the-problem-with-passwords/.
[139] Groeneveld, Rachid (2022). "The Password Problem." Nomios at https://www.nomios.com/news-blog/password-problem/.
[140] List, *supra*
[141] *Id.*

inadequate security practices. However, this continuous barrage of security alerts and password change requirements can lead to a sense of resignation and apathy, fostering complacency rather than encouraging proactive security measures.[142]

In light of these challenges, it is imperative for web3 companies to recognize the human factors underlying password memorability issues and implement user-friendly strategies to address them. From offering password management tools and mnemonic techniques to providing clear guidance on creating and safeguarding strong passwords, organizations can empower users to adopt more secure password practices without unduly burdening them with undue cognitive strain.

Public Keys

In the context of cryptographic systems, public keys are intended to be widely disseminated and freely shared, serving as essential components of asymmetric encryption schemes and digital signature mechanisms.[143] However, there are scenarios in which it may be prudent or necessary to keep public keys private, despite their name suggesting otherwise.[144] One such scenario involves protecting the integrity and confidentiality of communication channels in environments where adversaries may attempt to intercept or manipulate transmitted data.

---

[142] Groeneveld, *supra*.
[143] Staff (2024). "Public Key Encryption." Bugcrowd at https://www.bugcrowd.com/glossary/public-key-encryption/.
[144] Stanford at https://cs.stanford.edu/people/eroberts/courses/soco/projects/public-key-cryptography/probs.html.

By keeping public keys private in certain contexts, organizations can mitigate the risk of unauthorized entities exploiting the information to compromise the security of cryptographic protocols or gain illicit access to sensitive information.[145] For instance, in the realm of secure communication protocols such as Transport Layer Security (TLS), where public keys are used to facilitate encrypted data transmission between clients and servers, disclosing public keys to potential attackers could enable them to mount man-in-the-middle attacks or impersonate legitimate entities.[146]

Furthermore, in scenarios where the authenticity of public keys is crucial for establishing trust relationships between entities, maintaining strict control over the distribution and dissemination of public keys becomes paramount.[147] Any unauthorized modification or tampering with public keys could undermine the integrity of cryptographic operations and erode trust in the security of the underlying system.[148] Therefore, by treating public keys as sensitive assets and exercising judicious discretion in their handling and disclosure, organizations can uphold the confidentiality, integrity, and authenticity of cryptographic communications.

While the name "public key" implies openness and transparency, there are circumstances in which keeping public keys private is essential for safeguarding the security and trustworthiness of cryptographic systems.[149] By

---

[145] *Id.*
[146] Slubbs, Rob (2022). "Cryptographic Key Management- The Risks and Mitigation." Cryptomathic at https://www.cryptomathic.com/news-events/blog/cryptographic-key-management-the-risks-and-mitigations.
[147] *Id.*
[148] *Id.*

adopting a risk-based approach to public key management and implementing robust security measures to protect against unauthorized disclosure or tampering, organizations can uphold the confidentiality, integrity, and authenticity of cryptographic operations in diverse and challenging environments.[150]

Private Keys

Private keys play a critical role in cryptographic systems, serving as the cornerstone of digital signatures, encryption, and authentication mechanisms. Unlike public keys, which are intended for widespread dissemination, private keys must be kept confidential and securely guarded at all times. The security of private keys directly impacts the integrity and confidentiality of cryptographic operations, making their protection a paramount concern for individuals and organizations alike.[151]

One compelling reason to keep private keys secure is to prevent unauthorized access to sensitive data and assets controlled by cryptographic systems. Private keys are used to digitally sign transactions, decrypt encrypted data, and authenticate users in various contexts, including blockchain networks, secure messaging platforms, and online banking systems.[152] If private keys fall into the wrong hands due to negligence or inadequate security measures, malicious actors could compromise the security and privacy of sensitive information, leading to financial losses, identity theft, or unauthorized access to critical systems.

---

[149] Information Security at https://security.stackexchange.com/questions/54353/how-dangerous-might-be-publishing-a-public-key.

[150] Stanford, *supra.*

[151] Slubbs, *supra.*

[152] *Id.*

Moreover, the loss or compromise of private keys can have far-reaching consequences, including irreversible damage to digital identities, reputational harm, and legal liabilities. In blockchain ecosystems, for example, private keys are used to access and manage cryptocurrency assets stored in digital wallets. A breach of private key security could result in the unauthorized transfer or theft of funds, with little to no recourse for affected individuals or entities. Therefore, safeguarding private keys against theft, loss, or unauthorized disclosure is essential for preserving the confidentiality, integrity, and availability of cryptographic assets and resources.

Storage Software

Storage software for passwords and keys plays a crucial role in managing access to sensitive information and cryptographic assets across various digital platforms and applications.[153] However, despite its utility, such software can introduce significant security vulnerabilities if not implemented and configured properly. From inadequate encryption mechanisms to susceptibility to hacking attacks, several factors contribute to the inherent risks associated with storage software, underscoring the need for robust security measures and best practices.[154]

One of the primary concerns with storage software is the risk of data breaches resulting from insufficient encryption or poor security controls.[155] Passwords and keys stored in

---

[153] Valiaugaite, Inga (2024). "Are Password Managers Secure?" Cybernews at https://cybernews.com/best-password-managers/are-password-managers-safe/.
[154] Team (2022). "Six of the Biggest Problems with Password Managers." Hypr at https://blog.hypr.com/problems-with-password-managers.
[155] Robert (2023). "Why Password Managers Don't Solve the Password

plaintext or using weak encryption algorithms are susceptible to unauthorized access and interception by malicious actors. Without proper encryption and cryptographic protection, sensitive information becomes vulnerable to eavesdropping, interception, and tampering, compromising the confidentiality and integrity of stored credentials and cryptographic assets.

Furthermore, the centralized storage of passwords and keys in software-based repositories presents a single point of failure and an attractive target for cyberattacks.[156] Attackers may exploit vulnerabilities in storage software, such as software bugs, misconfigurations, or inadequate access controls, to gain unauthorized access to stored credentials and cryptographic keys. Once inside the system, threat actors can exfiltrate sensitive data, escalate privileges, or launch further attacks, posing significant risks to the security and integrity of digital assets and systems.

Moreover, the proliferation of cloud-based storage solutions and password management tools introduces additional security challenges and considerations.[157] While these platforms offer convenience and accessibility, they also introduce new attack vectors and dependencies on third-party service providers. Organizations must carefully evaluate the security features, data protection mechanisms, and compliance standards of storage software providers to mitigate the risks of data breaches, unauthorized access, and regulatory non-compliance.[158]

Problem." Corbado at https://www.corbado.com/blog/password-managers-problem.
[156] Valiaugaite, *supra*
[157] *Id.*
[158] Corbado, *supra.*

Another issue with storage software is the potential for insider threats and unauthorized access by privileged users or administrators. Even with robust access controls and encryption mechanisms in place, insider threats pose a significant risk to the security of stored passwords and keys.[159] Malicious insiders or compromised accounts may abuse their privileges to access, modify, or exfiltrate sensitive information, bypassing traditional security measures and controls.

Additionally, the lack of proper lifecycle management and auditing capabilities in storage software can hinder organizations' ability to track and monitor access to passwords and keys effectively. Without comprehensive audit trails and logging mechanisms, detecting unauthorized access or suspicious activities becomes challenging, increasing the likelihood of undetected security breaches and data leaks.

Furthermore, the reliance on storage software for password management and key storage introduces dependencies on proprietary or closed-source solutions, limiting transparency, and accountability.[160] Organizations may struggle to verify the security claims and assurances provided by software vendors, exacerbating trust issues and uncertainties regarding the integrity and resilience of stored credentials and cryptographic assets.

While storage software for passwords and keys offers convenience and efficiency in managing access to sensitive information and cryptographic assets, it also introduces significant security risks and challenges.[161] Organizations

---

[159] Valiaugaite, *supra*.
[160] Hypr, *supra*.
[161] Corbado, *supra*.

must implement robust encryption mechanisms, access controls, and auditing capabilities to safeguard stored credentials and cryptographic keys effectively. Additionally, regular security assessments, vulnerability scans, and penetration testing can help identify and mitigate potential vulnerabilities and weaknesses in storage software, ensuring the confidentiality, integrity, and availability of sensitive data and assets.

Dual Factor Problems (Using Same Device)

Using the same device for both accessing data and serving as a two-factor authentication (2FA) method introduces several inherent risks and vulnerabilities that can compromise the security of sensitive information and authentication mechanisms.[162] While 2FA is intended to enhance security by requiring users to provide two forms of verification, relying on the same device for both factors undermines the effectiveness of this security measure and exposes users to potential threats and attacks.

Firstly, using a single device for accessing data and 2FA increases the risk of a single point of failure. If the device is lost, stolen, or compromised, attackers gain access not only to the user's data but also to the secondary authentication factor, effectively bypassing the intended security controls.[163] This scenario nullifies the protective benefits of 2FA, as attackers can exploit the compromised device to impersonate legitimate users and gain unauthorized access to sensitive accounts and information.[164]

[162] Vaughan-Nichols, Steven (2022). "The Trouble with 2fa." CSO at https://www.csoonline.com/article/573739/the-trouble-with-2fa.html.
[163] Staff (2020). "Whats Wrong with Tw0-Factor Authentication." Security Scorecard at https://securityscorecard.com/blog/whats-wrong-with-two-factor-authentication/.
[164] *Id.*

Moreover, the convergence of data access and 2FA on the same device introduces a dependency that can be exploited by sophisticated adversaries.[165] Malicious actors may target the device itself, exploiting vulnerabilities or weaknesses to compromise its integrity and undermine the security of both data access and authentication processes.[166] By compromising the device, attackers can intercept or manipulate authentication signals, circumventing 2FA protections and gaining unauthorized access to protected resources.

Additionally, using the same device for data access and 2FA increases the attack surface and exposure to various security threats and malware.[167] If the device becomes infected with malicious software, such as keyloggers or spyware, attackers can capture sensitive information, including passwords, authentication codes, and other credentials, compromising both data confidentiality and authentication security.[168] Furthermore, malware may tamper with the 2FA process, intercepting authentication requests or altering authentication responses to facilitate unauthorized access.[169]

Furthermore, the dual use of the same device for data access and 2FA undermines the principle of segregation of duties, a fundamental security principle aimed at reducing the risk of insider threats and unauthorized access.[170] By

---

[165] Vaughan-Nichols, *supra*.
[166] Twitter Support at https://help.twitter.com/en/forms/account-access/regain-access/2fa-problem.
[167] *Id.*
[168] Vaughan-Nichols, *supra*.
[169] Wallen, Jack (2022). "Why 2FA is failing and what should be done about it." Tech Republic at https://www.techrepublic.com/article/why-2fa-is-failing/.

combining these functions on a single device, organizations fail to enforce proper separation between roles and responsibilities, increasing the likelihood of privilege misuse, conflicts of interest, and insider abuse. This lack of segregation exacerbates security risks and weakens the overall security posture of the organization.[171]

Moreover, using the same device for data access and 2FA may hinder the implementation of security best practices, such as device encryption, remote wipe capabilities, and multi-factor authentication (MFA) policies.[172] Organizations may struggle to enforce stringent security controls and policies on devices that serve dual roles, compromising the effectiveness of security measures and exposing sensitive data to greater risk of unauthorized access, interception, or theft.

One of the greatest weaknesses of 2FA is relying on the same device for both accessing data and serving as a 2FA method.[173] This introduces significant security risks and vulnerabilities that can compromise the confidentiality, integrity, and availability of sensitive information and authentication mechanisms.[174] Organizations must adopt a risk-based approach to security, implementing robust controls, segregation of duties, and security best practices to mitigate the risks associated with dual-use devices and ensure the effectiveness of their security measures.

---

[170] *Id.*

[171] Vaughan-Nichols, *supra.*

[172] *Id.*

[173] Elgan, Mike (2021). "How to Fix the Big Problem with two factor and Multi-Factor Authentication." Security Intelligence at https://securityintelligence.com/articles/fix-big-problems-with-two-factor-multifactor-authentication/.

[174] *Id.*

Conclusion

In conclusion, the challenges surrounding password storage and authentication within Web3 companies highlight the critical need for robust security measures and practices to safeguard sensitive information and protect against unauthorized access. From the difficulty of remembering complex passwords to the vulnerabilities associated with storing public and private keys, organizations must address various security concerns to mitigate risks effectively.

Furthermore, the reliance on storage software for passwords and keys underscores the importance of implementing secure and resilient solutions that prioritize data protection and confidentiality. With the increasing sophistication of cyber threats and attacks, organizations must remain vigilant and proactive in securing their authentication mechanisms and safeguarding access to critical assets and resources.

Moreover, the pitfalls of using the same device for both data access and two-factor authentication emphasize the need for segregation of duties and the implementation of robust security controls to mitigate the risks associated with dual-use devices. By adopting a risk-based approach to security and implementing multi-layered defenses, organizations can enhance their security posture and better defend against evolving threats and vulnerabilities.

Overall, addressing the complexities of password storage and authentication in Web3 environments requires a comprehensive and proactive security strategy that integrates best practices, technologies, and policies. By prioritizing security, organizations can effectively mitigate risks, protect sensitive data, and maintain the trust and confidence of their stakeholders in an increasingly interconnected and digital world.

Chapter 6: Data Encryption

"You know the interesting thing about encryption is that it canot be secure for just some people."

Pavel Durov[175]

Chapter six breaks down the intricate challenges and complexities surrounding data encryption within the realm of Web3, highlighting the pivotal role encryption plays in safeguarding sensitive information while also shedding light on the various issues and considerations that arise in its implementation. The chapter navigates through key areas such as key management, algorithm selection and implementation, legacy integration, scalability and performance concerns, and the looming specter of quantum computing, offering insights into the nuances of encryption practices in modern digital environments.

Key management emerges as a central concern, with organizations grappling with the complexities associated with securely generating, storing, and distributing encryption keys. The sheer complexity of managing cryptographic keys, coupled with the dichotomy between centralization and decentralization approaches, presents formidable challenges, further compounded by the inherent susceptibility to human error, underscoring the critical need for robust key management strategies.

Algorithm selection and implementation represent another crucial aspect of data encryption, with organizations tasked with choosing efficient and robust algorithms while mitigating the risks of human flaws in programming or

[175] (2024). "Encryption Quotes, Pavel Durov." Brainy Quotes at https://www.brainyquote.com/topics/encryption-quotes.

implementation. The chapter underscores the importance of staying abreast of advancements in encryption technology while remaining vigilant against emerging threats and vulnerabilities.

Legacy integration poses significant hurdles, with interoperability problems, workarounds, and standardization issues complicating the seamless integration of encryption protocols across diverse systems and platforms. The chapter delves into the intricacies of legacy systems and the imperative for organizations to navigate the integration process diligently to ensure data security and operational efficiency.

Scalability and performance considerations loom large in the encryption landscape, with organizations grappling with the tradeoffs between security and efficiency. Balancing the need for robust encryption with the demands of resource-intensive data systems presents a formidable challenge, necessitating careful planning and strategic decision-making.

Finally, the advent of quantum computing heralds both promises and perils for data encryption, with quantum attacks posing existential threats to conventional cryptographic methods. The chapter explores the concept of quantum-resistant algorithms and emphasizes the importance of future-proofing encryption strategies to withstand the disruptive forces of quantum computing, advocating for a proactive approach to evolving encryption practices that grow with the system.

In essence, this chapter elucidates the multifaceted nature of data encryption challenges in the Web3 landscape, offering insights into the complexities of key management, algorithm selection, legacy integration, scalability, and the looming specter of quantum computing. By addressing these challenges head-on and adopting a proactive stance towards encryption practices, organizations can fortify their data security posture and navigate the evolving threat landscape with confidence.

Key Management

Key management is an important part of security within Web3 companies, playing a pivotal role in safeguarding sensitive information and ensuring the integrity and confidentiality of data transmissions.[176] At its essence, effective key management involves the secure generation, storage, distribution, and revocation of cryptographic keys that underpin encryption protocols.[177] The vitality of robust key management practices lies in its ability to thwart unauthorized access, mitigate the risk of data breaches, and uphold the trust and confidentiality essential in digital transactions.

First and foremost, robust key management is vital for protecting sensitive information from unauthorized access and malicious adversaries.[178] Encryption keys serve as the linchpin of cryptographic algorithms, rendering data indecipherable to unauthorized parties without the corresponding keys.[179] By implementing stringent key

---

[176] Saha, Parnashree (2022). "10 Enterprise Encryption Key Management Best Practices." Encryption Consulting at https://www.encryptionconsulting.com/10-enterprise-encryption-key-management-best-practices/.
[177] *Id.*
[178] Federal Partnerships for Interoperable communications at https://www.cisa.gov/sites/default/files/2023-02/08-19-2020_Operational-Best-Practices-for-Encryption-Key-Mgmt_508c.pdf.
[179] *Id.*

management protocols, Web3 companies can mitigate the risk of data interception, eavesdropping, and tampering, thereby safeguarding the confidentiality and integrity of their digital assets.

Furthermore, effective key management is essential for ensuring compliance with regulatory mandates and industry standards governing data protection and privacy.[180] Many jurisdictions impose stringent requirements on the protection of sensitive information, mandating the use of encryption and robust key management practices to safeguard data against unauthorized access and disclosure. Failure to implement adequate key management measures can expose companies to regulatory scrutiny, legal liabilities, and reputational damage.[181]

Moreover, robust key management enhances operational resilience by safeguarding against the risk of key loss or compromise.[182] Loss of encryption keys or unauthorized access to key materials can render encrypted data inaccessible, disrupting business operations and potentially leading to data loss or system downtime. By instituting robust key management protocols, Web3 companies can mitigate the risk of key-related incidents and ensure the continuous availability and accessibility of encrypted data.[183]

Additionally, effective key management fosters trust and confidence among stakeholders, including customers,

---

[180] Pal, Kaushik (2023). "10 Best Practices for Encryption Key management and Data Security." Techopedia at https://www.techopedia.com/2/30767/security/10-best-practices-for-encryption-key-management-and-data-security.
[181] Saha, *supra.*
[182] *Id.*
[183] CISA, *supra.*

partners, and investors. In an era marked by escalating cyber threats and data breaches, demonstrating a commitment to robust key management practices signals a company's dedication to safeguarding sensitive information and preserving the confidentiality and privacy of customer data.[184] This, in turn, enhances brand reputation, fosters customer loyalty, and cultivates a competitive edge in the marketplace.

Ultimately, key management serves as the linchpin of security within Web3 companies, underpinning the integrity, confidentiality, and resilience of digital transactions and communications.[185] By prioritizing robust key management practices, organizations can fortify their defenses against evolving cyber threats, comply with regulatory mandates, enhance operational resilience, and instill trust and confidence among stakeholders. As Web3 ecosystems continue to evolve, the imperative for effective key management remains paramount in safeguarding the digital assets and preserving the trust of users in the digital realm.

Algorithm Selection and Implementation

Algorithm selection and implementation play a pivotal role in safeguarding data encryption within Web3 companies, dictating the effectiveness, resilience, and security posture of cryptographic protocols.[186] At the heart of this endeavor lies the quest for efficient and robust algorithms capable of

---

[184] OWASP at https://cheatsheetseries.owasp.org/cheatsheets/Key_Management_Cheat_Sheet.html.
[185] Pal, *supra*.
[186] O'Neil, James (2009) cognitive Radio Performance Analysis at https://www.sciencedirect.com/topics/computer-science/algorithm-selection.

withstanding adversarial attacks, preserving data integrity, and ensuring confidentiality across digital transactions and communications.

First and foremost, the selection of efficient and robust algorithms is paramount to the security of data encryption in Web3 companies. Robust algorithms are characterized by their ability to withstand sophisticated cryptographic attacks and maintain data confidentiality even in the face of concerted adversarial efforts.[187] Moreover, efficiency is essential to optimize computational resources and minimize processing overhead, ensuring seamless integration with existing infrastructure and enabling timely execution of cryptographic operations.

However, the efficacy of cryptographic algorithms can be compromised by human flaws in programming or implementation, underscoring the importance of rigorous software development practices and adherence to industry best practices.[188] Human errors, such as coding vulnerabilities, implementation flaws, or misconfigurations, can inadvertently expose cryptographic systems to exploitation by malicious actors, undermining their integrity and rendering them susceptible to compromise. To mitigate this risk, Web3 companies must prioritize security-conscious development practices, conduct thorough code reviews, and employ robust testing methodologies to identify and rectify potential vulnerabilities before deployment.

Furthermore, vigilant monitoring of advancements in cryptographic research and technology is imperative to stay

---

[187] Staff (2022). "Algorithm Selection for Machine Learning." Elite Data Science at https://elitedatascience.com/algorithm-selection.
[188] O'Neil, *supra.*

abreast of emerging threats and evolving security requirements. As cryptographic algorithms and protocols continue to evolve in response to emerging threats and technological advancements, Web3 companies must remain vigilant and proactive in evaluating and adopting cutting-edge encryption techniques that offer enhanced security guarantees and resilience against emerging attack vectors.

Moreover, the selection and implementation of cryptographic algorithms must be guided by a holistic approach that considers not only their security properties but also their interoperability, scalability, and performance characteristics. Interoperability ensures seamless integration with existing systems and technologies, enabling interoperable data exchange and communication across diverse platforms and ecosystems. Scalability is crucial to accommodate the growing volume and complexity of digital transactions and communications, ensuring that cryptographic protocols can scale efficiently to meet evolving business needs without compromising security or performance.[189]

Ultimately, algorithm selection and implementation represent a critical aspect of data encryption in Web3 companies, shaping the resilience, effectiveness, and security posture of cryptographic systems. By prioritizing the adoption of efficient and robust algorithms, addressing human flaws in programming or implementation, and staying vigilant to advancements in cryptographic research and technology, organizations can enhance the security and integrity of their digital assets, safeguard sensitive information, and maintain trust and confidence among stakeholders in the digital realm.

---

[189] *Id.*

Legacy Integration

Legacy equipment and coding integration present significant challenges to maintaining robust encryption standards within a company's infrastructure, often leading to interoperability problems, necessitating workarounds, and exacerbating standardization issues.[190]

Interoperability problems arise when legacy systems fail to seamlessly integrate with modern encryption standards, hindering data exchange and communication between disparate platforms and technologies.[191] Incompatibilities between legacy equipment and contemporary encryption protocols can result in data silos, communication bottlenecks, and operational inefficiencies, undermining the organization's ability to leverage encryption effectively across its digital ecosystem.[192]

To circumvent interoperability challenges, organizations may resort to implementing ad-hoc workarounds or interim solutions to bridge the gap between legacy and modern encryption standards.[193] However, these makeshift solutions often introduce additional complexity, security vulnerabilities, and maintenance overhead, further complicating the encryption landscape and increasing the organization's exposure to cyber threats and compliance risks.[194]

---

[190] Davis, Angela (2023). "Deep-Diving into Legacy Systems Integration." Open Legacy at https://www.openlegacy.com/blog/legacy-systems-integration.
[191] Dilmegani, Cem (2024). "Legacy System Integration in 2024: Top 4 Use Cases." AI Multiple at https://research.aimultiple.com/legacy-system-integration/.
[192] *Id.*
[193] Zheldak, Pavlo (2023). "How to Integrate Legacy Systems and Modern Software." Acropolium at https://acropolium.com/blog/legacy-systems-integration/.

Standardization issues can impede the adoption and deployment of encryption standards across the organization's infrastructure.[195] This can hamper efforts to establish consistent encryption policies, procedures, and practices.[196] Inconsistent implementation of encryption standards due to legacy equipment and coding integration can create fragmentation, ambiguity, and confusion regarding encryption protocols, weakening the organization's overall security posture and undermining its ability to enforce uniform encryption practices across different departments and business units.[197]

Addressing legacy equipment and coding integration challenges requires a strategic approach that prioritizes interoperability, compatibility, and standardization across the organization's digital infrastructure.[198] By investing in modernization initiatives, legacy systems can be upgraded or replaced with interoperable solutions that seamlessly integrate with contemporary encryption standards, enabling seamless data exchange, communication, and collaboration across the organization's digital ecosystem. Additionally, organizations must adopt standardized encryption protocols and best practices to ensure consistency, reliability, and compliance with regulatory requirements, thereby enhancing data security, protecting sensitive information,

---

194 Gitlin, Jon (2023). "What is a Legacy System? And How can you integrate one with modern apps." Workato at https://www.workato.com/the-connector/what-is-legacy-system/.
195 Davis, *supra*.
196 Dilmegani, *supra*.
197 *Id.*
198 Staff (2024). "How to Integrate Legacy Systems and Modern SaaS Applications without Disruption." Cleo at https://www.cleo.com/blog/integrate-legacy-systems-saas-applications.

and mitigating cyber risks in an increasingly interconnected and digitally driven world.[199]

Scalability and Performance

Scalability and performance issues pose significant challenges to encryption within web3 companies, often stemming from the need to balance security requirements with operational efficiency and resource constraints.[200] One common issue is the tendency to oversimplify encryption protocols for the sake of speed, compromising security in favor of performance gains.[201] While faster encryption algorithms may improve data processing speed, they often sacrifice cryptographic strength and resilience, leaving sensitive information vulnerable to exploitation by malicious actors.[202]

Resource-intensive data systems exacerbate scalability and performance challenges, especially in web3 environments where large volumes of data are processed and transmitted in real-time.[203] Encryption adds an additional layer of computational overhead, requiring significant processing power and memory resources to encrypt and decrypt data efficiently.[204] As a result, resource-constrained systems

---

[199] Davis, *supra*.
[200] Peter, Sinu (2019). "6 Ways Scalability Requirements are Impacting Today's Cyber Security." Cyber Proof at https://blog.cyberproof.com/blog/6-ways-scalability-requirements-are-impacting-todays-cybersecurity.
[201] Bridgewater, Adrian (2022). "What does IT Scalability Mean?" Forbes at https://www.forbes.com/sites/adrianbridgwater/2022/07/04/what-does-it-scalability-actually-mean/?sh=2863fdf43429.
[202] Peter, *supra*.
[203] *Id.*
[204] Staff (2023). "The Challenges of Software Scalability and How to Overcome Them." Pilotcore at https://pilotcoresystems.com/insights/the-challenges-of-software-

may experience bottlenecks, latency issues, and performance degradation when handling encrypted data, impacting user experience, application responsiveness, and overall system reliability.

Moreover, there exists a delicate balance between security and efficiency in encryption, as stronger cryptographic algorithms typically incur higher computational costs and resource utilization.[205] Organizations must carefully weigh the tradeoffs between security and performance when designing encryption solutions, ensuring that encryption overhead does not compromise system scalability, responsiveness, or throughput.[206] However, prioritizing efficiency over security may expose the organization to increased cyber risks, data breaches, and compliance violations, underscoring the importance of striking a balance between security requirements and operational demands.

One approach to mitigating scalability and performance issues in encryption is to adopt scalable and efficient encryption algorithms that strike an optimal balance between cryptographic strength and computational efficiency.[207] By leveraging advanced encryption techniques such as elliptic curve cryptography (ECC) and homomorphic encryption, organizations can achieve robust security without sacrificing performance, enabling them to

---

scalability-and-how-to-overcome-them/.

[205] Bridgewater, *supra*.

[206] *Id.*

[207] Staff (2023). "Top 10 Barriers in Digital Product Engineering: Overcoming Scalability Challenges." UST at https://www.ust.com/en/insights/top-10-barriers-in-digital-product-engineering-overcoming-scalability-challenges.

scale encryption operations across distributed systems and high-volume data streams effectively.[208]

Additionally, organizations can optimize encryption performance by implementing parallel processing techniques, hardware acceleration, and distributed computing architectures tailored to handle encryption tasks efficiently.[209] By offloading encryption operations to specialized hardware or leveraging cloud-based encryption services, organizations can alleviate computational burdens, improve scalability, and enhance overall system performance without compromising security.

Ultimately, addressing scalability and performance issues in encryption requires a holistic approach that integrates advanced cryptographic techniques, optimized algorithms, and scalable infrastructure to meet the evolving security and operational needs of web3 companies.[210] By adopting a proactive stance towards encryption optimization and performance tuning, organizations can enhance their resilience to cyber threats, safeguard sensitive data, and maintain robust encryption practices in a dynamic and demanding digital landscape.

Quantum Computing

Quantum computing represents a significant threat to encryption protocols widely used today, posing a formidable challenge to the security of web3 companies and their sensitive data.[211] Unlike classical computers, which rely on binary bits to perform calculations, quantum

---

[208] *Id.*
[209] Bridgewateer, *supra*.
[210] *Id.*
[211] Fisher, KAG et al (2018). "Quantum Computing and Encrypted Data." At https://arxiv.org/pdf/1309.2586.pdf.

computers leverage quantum bits or qubits, which can exist in multiple states simultaneously.[212] This inherent parallelism enables quantum computers to solve complex mathematical problems exponentially faster than classical computers, including those underlying widely-used encryption algorithms such as RSA and ECC.[213]

One of the most concerning implications of quantum computing for encryption is its potential to render existing cryptographic schemes obsolete by effectively breaking them through quantum algorithms like Shor's algorithm. These algorithms exploit quantum properties to factor large numbers and compute discrete logarithms—the foundation of many encryption schemes—significantly faster than classical computers.[214] As a result, quantum adversaries could decrypt encrypted communications, compromise digital signatures, and undermine the confidentiality and integrity of sensitive data protected by current encryption standards.[215]

To mitigate the threat posed by quantum computing, web3 companies must adopt quantum-resistant encryption algorithms that remain secure even in the presence of quantum adversaries.[216] These quantum-resistant algorithms, also known as post-quantum cryptography or quantum-safe cryptography, are designed to withstand

---

[212] Kirsch, Zach (2015). "Computer Systems Security." Tuffs University at http://www.cs.tufts.edu/comp/116/archive/fall2015/zkirsch.pdf.
[213] *Id.*
[214] Mavroeidis, Vasileios (2018). "The Impact of Quantum Computing on Present Cryptography." International Journal of Advanced Computer Science and Application 9:3 at https://arxiv.org/pdf/1804.00200.pdf.
[215] Fisher, *supra*.
[216] *Id.*

attacks from both classical and quantum adversaries, offering robust security guarantees in the quantum era.[217] Examples of quantum-resistant algorithms include lattice-based cryptography, hash-based signatures, and code-based cryptography, which leverage mathematical problems believed to be hard for both classical and quantum computers to solve.[218]

By proactively transitioning to quantum-resistant encryption algorithms, web3 companies can future-proof their security infrastructure and protect their sensitive data against emerging quantum threats.[219] This strategic approach ensures that encrypted communications, digital signatures, and data storage remain secure and resilient in the face of rapid advancements in quantum computing technology. Moreover, deploying quantum-resistant encryption algorithms today demonstrates a commitment to long-term security and helps organizations stay ahead of the curve by anticipating and mitigating future risks posed by quantum adversaries.[220]

In addition to adopting quantum-resistant encryption algorithms, web3 companies should also invest in research and development efforts to explore and deploy quantum-enhanced security solutions that leverage the unique capabilities of quantum computing to strengthen

---

[217] *Id.*

[218] Castelvecchi, Davide (2023). "Are Quantum Computers about to break online privacy?" Scientific American at https://www.scientificamerican.com/article/are-quantum-computers-about-to-break-online-privacy/.

[219] *Id.*

[220] Montague, Zach (2023). "The Race to Save Our Secrets from the Computers of the Future." The New York Times at https://www.nytimes.com/2023/10/22/us/politics/quantum-computing-encryption.html.

cybersecurity defenses.[221] Quantum key distribution (QKD), for example, enables the secure exchange of cryptographic keys using the principles of quantum mechanics, offering unprecedented levels of security for key distribution and encryption key management.[222]

Overall, the looming threat of quantum computing underscores the importance of proactive cybersecurity measures and strategic investments in quantum-resistant encryption technologies. By embracing quantum-safe cryptography, web3 companies can fortify their defenses against quantum threats, safeguard their digital assets, and maintain trust and confidence in their security posture amidst the disruptive forces of quantum technology.

Conclusion

In conclusion, the challenges associated with data encryption in web3 companies are multifaceted and demand careful consideration to ensure robust security measures. Key management emerges as a pivotal aspect of encryption, as the proper generation, storage, and protection of cryptographic keys are fundamental to the integrity and confidentiality of encrypted data. Moreover, algorithm selection and implementation play a critical role in safeguarding against potential vulnerabilities and exploits, highlighting the importance of leveraging efficient and robust encryption algorithms while remaining vigilant against human errors and emerging threats.

Legacy integration poses significant obstacles to encryption standards within web3 companies, as interoperability issues, workarounds, and standardization concerns can

[221] Kirsch, *supra*

[222] Castelvecchi, *supra*.

compromise the effectiveness and reliability of encryption mechanisms. Addressing these challenges requires proactive measures to update and modernize encryption protocols, streamline integration processes, and adhere to industry best practices to ensure seamless interoperability and compatibility across diverse systems and platforms.

Scalability and performance considerations present additional complexities in encryption, as organizations must balance the need for speed and efficiency with robust security measures. Over-simplification for speed can inadvertently weaken encryption defenses, while resource-intensive data systems may strain computational resources and compromise overall system performance. Striking the right balance between security and efficiency is paramount, necessitating careful optimization and fine-tuning of encryption protocols to meet the evolving needs of web3 companies.

The looming threat of quantum computing underscores the importance of adopting quantum-resistant encryption algorithms and exploring quantum-enhanced security solutions to mitigate the potential risks posed by quantum adversaries. By embracing quantum-safe cryptography and investing in quantum-enhanced security technologies, web3 companies can future-proof their encryption infrastructure and stay ahead of emerging threats in the rapidly evolving cybersecurity landscape.

Addressing the challenges of data encryption in web3 companies requires a holistic approach that encompasses key management, algorithm selection, legacy integration, scalability, performance optimization, and quantum resilience. By prioritizing security, staying informed about emerging threats, and implementing proactive measures to strengthen encryption defenses, organizations can effectively safeguard their data assets and uphold the trust and integrity of their digital operations in the web3 era.

Chapter 7: Updating Software

"People who are really serious about softeware should make their own hardware."

Alan Kay[223]

The failure to update software can pave the way for a host of security vulnerabilities and expose organizations to significant risks. This chapter looks into this critical aspect of cybersecurity, shedding light on the repercussions of neglecting software updates and the proactive measures that can mitigate potential threats. With a structured approach encompassing insights on the evolution of software vulnerabilities, the importance of security updates, and the efficacy of whitelists, this chapter serves as a comprehensive guide for safeguarding digital infrastructure against exploitation and intrusion.

The chapter begins by tracing the trajectory from seemingly innocuous software glitches to catastrophic security breaches, exemplified by the transition from the infamous "Blue Screen of Death" to sophisticated cyberattacks that leverage unpatched vulnerabilities. Through a retrospective lens, readers gain a nuanced understanding of how vulnerabilities in software systems have evolved over time, underscoring the imperative for organizations to remain vigilant and proactive in addressing potential points of weakness.

Central to the discourse on software security is the notion of security updates, which represent a cornerstone of modern cybersecurity practices. By promptly applying

---

[223] (2024). "Software Quotes." Brainy Quote at https://www.brainyquote.com/topics/software-quotes.

patches and updates released by software vendors, organizations can preemptively address known vulnerabilities and fortify their defense mechanisms against emerging threats. However, the efficacy of security updates hinges on timely implementation and robust change management processes, highlighting the importance of proactive risk management strategies in safeguarding digital assets.

In addition to security updates, the chapter delves into the concept of whitelists as a proactive measure for enhancing software security. By establishing a curated list of approved applications and processes, organizations can restrict access to authorized entities while mitigating the risks associated with untrusted or malicious software. Through the judicious implementation of whitelisting policies, organizations can bolster their security posture and minimize the likelihood of unauthorized access or compromise.

The insights presented in this chapter underscore the critical importance of proactive software maintenance and security hygiene in mitigating the risks posed by software vulnerabilities. By embracing a proactive approach to software updates, organizations can effectively mitigate the risks of exploitation and enhance their resilience to evolving cyber threats. Furthermore, the adoption of whitelisting mechanisms serves as a complementary strategy for fortifying digital defenses and safeguarding sensitive assets against unauthorized access or compromise. Through a holistic approach to software security, organizations can navigate the complex terrain of cybersecurity with confidence and resilience.

From Blue Screen of Death to Major Hacks

The evolution of computer problems from the infamous "Blue Screen of Death" to modern-day cyber hacks stands as a testament to the ever-changing landscape of software vulnerabilities and the escalating sophistication of cyber threats. Initially, the Blue Screen of Death—an error screen displayed by Microsoft Windows operating systems—was emblematic of system crashes resulting from hardware or software malfunctions, often attributed to compatibility issues, device driver errors, or system file corruption. While disruptive, these incidents were predominantly viewed as technical glitches rather than deliberate security breaches.

However, as technology advanced and computing systems became more interconnected, the vulnerabilities inherent in software systems began to attract the attention of malicious actors seeking to exploit weaknesses for personal gain or malicious intent.[224] Over time, the paradigm shifted from isolated system failures to orchestrated cyberattacks targeting software vulnerabilities, with the potential for far-reaching consequences. This evolution reflects a fundamental change in the nature of computer problems, transcending mere technical errors to encompass deliberate exploitation of security flaws for nefarious purposes.

The proliferation of unpatched vulnerabilities in software systems provided fertile ground for the emergence of sophisticated cyber threats, ranging from malware infections and ransomware attacks to data breaches and system compromises.[225] As organizations and individuals increasingly relied on interconnected digital infrastructure

224

225

for essential functions and services, the stakes grew higher, amplifying the potential impact of software-related security incidents.[226] Moreover, the interconnected nature of modern computing ecosystems facilitated the rapid spread of malware and exploits, exacerbating the severity of cyber threats on a global scale.

Today, the consequences of failing to address software vulnerabilities extend far beyond mere inconvenience or system instability, with cyberattacks posing significant risks to critical infrastructure, sensitive data, and organizational resilience.[227] From high-profile data breaches targeting multinational corporations to disruptive ransomware attacks crippling essential services, the repercussions of software-related security incidents underscore the urgent need for proactive risk mitigation strategies and robust cybersecurity measures.

In essence, the evolution of computer problems from the Blue Screen of Death to modern cyber hacks reflects the dynamic interplay between technological innovation and malicious intent, highlighting the imperative for organizations and individuals to remain vigilant and proactive in addressing software vulnerabilities.[228] By understanding the historical context and evolving nature of software-related security threats, stakeholders can better appreciate the importance of timely software updates, robust cybersecurity practices, and proactive risk management strategies in safeguarding digital assets and

---

[226] Staff (2024). "The Evolution of the Hacker." Techgaurd Security at https://blog.techguard.com/the-evolution-of-the-hacker.
[227] Tripwire (2016). "The Evolution of Hacking." Fortra at https://www.tripwire.com/state-of-security/the-evolution-of-hacking.
[228] Fox, Jacob (2022). "A Brief History of Hacking." Cobalt at https://www.cobalt.io/blog/history-of-hacking.

preserving operational continuity in an increasingly interconnected world.

Security Updates

Software security updates, also known as patches or fixes, are essential components of maintaining the integrity and security of computer systems, applications, and software products. These updates are designed to address identified vulnerabilities, bugs, or weaknesses in software code that could be exploited by malicious actors to compromise system security, steal sensitive data, or disrupt normal operations.[229] By releasing timely updates, software vendors and developers aim to remediate known security flaws, enhance system resilience, and minimize the risk of exploitation by cyber attackers.

One of the primary reasons why software security updates are crucial lies in their role in mitigating the ever-present threat of cyberattacks and data breaches.[230] As cyber threats continue to evolve and grow in sophistication, software vulnerabilities represent prime targets for exploitation by threat actors seeking unauthorized access to systems or sensitive information.[231] Security updates serve as a proactive defense mechanism against such threats, helping to close security gaps, strengthen system defenses, and reduce the likelihood of successful cyber intrusions.

---

[229] Staff (2023). "Why is it important to update security patches?" Network Coverage at https://www.netcov.com/why-is-it-important-to-update-security-patches/.

[230] Staff (2023). "What is a security patch and why is it important." Teal at https://tealtech.com/blog/it-services/cybersecurity/what-is-a-security-patch/.

[231] *Id.*

Software security updates play a critical role in promoting the principle of "defense in depth" within cybersecurity frameworks.[232] By continuously identifying and addressing vulnerabilities in software products, organizations can establish multiple layers of defense against potential security threats, thereby enhancing overall resilience and reducing the likelihood of successful attacks. This proactive approach to security emphasizes the importance of regular updates as part of a comprehensive cybersecurity strategy aimed at minimizing risk and safeguarding critical assets.

In addition to addressing security vulnerabilities, software updates often include enhancements, bug fixes, and performance optimizations that contribute to the overall reliability and functionality of software products.[233] By staying current with the latest updates, users can benefit from improved stability, usability, and compatibility, leading to a more seamless and efficient computing experience.[234] Furthermore, timely installation of updates helps to ensure ongoing support and maintenance of software applications, prolonging their lifespan and maximizing return on investment for organizations and end-users alike.

However, despite the clear benefits of software security updates, challenges related to patch management,

---

[232] Staff (2023). "What are security patch updates and why do they matter?" RSI Security at https://blog.rsisecurity.com/what-are-security-patch-updates-and-why-do-they-matter/.

[233] CISA (2023). "Understanding Patches and Software Updates." Cybersecurity & Infrastructure Security Agency at https://www.cisa.gov/news-events/news/understanding-patches-and-software-updates.

[234] Staff (2024). "Why it's important to install Windows updates and patches." Zunesis at https://www.zunesis.com/why-install-windows-updates/.

compatibility issues, and system disruptions can pose obstacles to their effective deployment and adoption. Organizations must carefully balance the need for security with operational considerations to minimize disruptions and maintain productivity during the update process.[235] Additionally, the increasing complexity of modern software ecosystems, coupled with the proliferation of interconnected devices and platforms, underscores the importance of implementing robust patch management practices and security protocols to effectively manage and mitigate software-related risks.

Whitelists

A whitelist is a security mechanism used to control access to resources or functionalities by explicitly specifying which entities or actions are permitted, while denying all others by default.[236] In the context of data security, whitelists are often employed to define a list of trusted entities, such as users, applications, or network addresses, that are authorized to interact with a system or access sensitive data. By restricting access to known, approved entities and configurations, whitelists help to prevent unauthorized access, reduce the attack surface, and mitigate the risk of security breaches or data exfiltration.

One of the key advantages of whitelists is their proactive approach to security, which focuses on allowing only pre-approved entities or actions to occur within a system or network environment.[237] Unlike traditional blacklisting approaches, which attempt to identify and block known

---

235 CISA, *supra*.
236 Web3auth at https://web3auth.io/docs/dashboard-setup/whitelisting.
237 Staff (2022) "Whitelist" Metaschool at https://metaschool.so/articles/whitelist-meaning/.

threats or malicious entities, whitelists operate on the principle of least privilege, granting access only to the specific resources or functionalities required for legitimate purposes.[238] By adhering to the principle of least privilege, organizations can minimize the potential for unauthorized access, privilege escalation, or misuse of system resources, thereby enhancing overall security posture and reducing the likelihood of successful cyber-attacks.

Whitelists offer organizations greater control and visibility over their digital environments, enabling them to enforce security policies, regulatory compliance requirements, and best practices for data protection.[239] By maintaining a centralized list of trusted entities or configurations, administrators can effectively manage access permissions, monitor system activity, and detect anomalous behavior or unauthorized access attempts in real-time. This granular control and visibility empower organizations to enforce stricter security measures, identify and remediate security risks proactively, and maintain the confidentiality, integrity, and availability of sensitive data assets.[240]

Conclusion

In conclusion, the evolution of computer problems from the era of the Blue Screen of Death to modern-day hacks highlights the critical importance of keeping software updated and secure. Neglecting to update software leaves

---

[238] Pierce, Michael (2023). "How to Establish a High security Whitelisting process in Web3." Spiceworks at https://www.spiceworks.com/tech/innovation/guest-article/secure-web3-whitelisting/.
[239] George, Alen (2024). "How to collect verified crypto wallet addresses from Whitelists, Allowlists, Waitlists and Airdrops." Block survey at https://blocksurvey.io/features/web3-identity-verification.
[240] Web3auth, *supra.*

systems vulnerable to exploitation by cybercriminals who exploit known vulnerabilities to launch attacks, compromise data, and disrupt operations. Security updates play a vital role in addressing these vulnerabilities by providing patches, fixes, and enhancements that mitigate the risk of exploitation and strengthen the overall security posture of systems and networks.

Furthermore, whitelists serve as an essential security control mechanism, enabling organizations to define and enforce strict access policies that restrict access to trusted entities and configurations. By implementing whitelists, organizations can reduce the attack surface, minimize the risk of unauthorized access, and enhance visibility and control over system activities. This proactive approach to security aligns with industry best practices and regulatory requirements, helping organizations to achieve compliance and safeguard sensitive data assets from unauthorized access or misuse.

In addition to security updates and whitelists, organizations must prioritize regular security assessments, monitoring, and incident response capabilities to detect and respond to emerging threats effectively. By adopting a comprehensive approach to cybersecurity, organizations can address vulnerabilities, mitigate risks, and protect their data and assets from evolving cyber threats. Ultimately, proactive measures such as software updates and whitelists, combined with robust security practices, are essential components of a resilient cybersecurity strategy that helps organizations stay ahead of threats and safeguard their digital assets.

Chapter 8: Incident Response and Public Relations

"Is hacking acceptable? It depends on the Motive."

Charlie Brooker[241]

This chapter discusses the intricate realm of incident response and public relations in the context of major security breaches within web3 companies. As organizations increasingly rely on digital infrastructure and technologies to drive their operations, the risk of security incidents and data breaches looms large, underscoring the critical importance of robust incident response strategies and effective public relations management. In today's interconnected and data-driven business landscape, the ability to swiftly and adeptly address security breaches while maintaining stakeholder trust and brand reputation is paramount for organizational resilience and longevity.

The inevitability of failure in any organization serves as a sobering reminder of the ever-present threat landscape and the need for proactive preparedness. Despite best efforts to fortify defenses and implement stringent security measures, no company is immune to security breaches and cyberattacks. Acknowledging this reality, organizations must adopt a proactive stance towards incident response, recognizing that it is not a matter of if but when a security incident will occur. By embracing this mindset and investing in comprehensive incident response protocols, organizations can better position themselves to detect, mitigate, and recover from security breaches effectively.

---

[241] (2024). "Hacking Quotes, Charlie Brooker." Brainy Quote at https://www.brainyquote.com/topics/hacking-quotes.

Responding quickly and decisively to security breaches is essential for containing the threat and minimizing the impact on business operations and stakeholders. The ability to rapidly identify, assess, and mitigate security incidents is paramount for preventing further escalation and limiting the extent of damage. Timely intervention enables organizations to isolate compromised systems, implement remediation measures, and restore normalcy, thereby reducing downtime and mitigating financial and reputational repercussions. Through prompt and coordinated action, organizations can demonstrate their commitment to cybersecurity and their ability to safeguard sensitive data and assets from malicious actors.

Fostering a culture of transparency, accountability, and collaboration within the organization is fundamental to effective incident response. Employees must feel empowered to report security incidents promptly without fear of retribution or blame, enabling organizations to address vulnerabilities and threats swiftly. By removing barriers to communication and encouraging open dialogue, organizations can leverage the collective expertise and insights of their workforce to strengthen their security posture and enhance their resilience against evolving cyber threats. Effective incident response is not solely a technical endeavor but also a cultural and organizational imperative that requires collaboration across departments and levels of hierarchy.

The Inevitability of Failure

In the dynamic landscape of modern business operations, the inevitability of failure is a harsh yet undeniable reality that all companies must confront.[242] No matter how robust

[242] Staff (2022). "3 Bottom-Line Reasons Why Every Business Should

or sophisticated a company's systems and processes may be, unforeseen challenges, vulnerabilities, and disruptions are bound to occur.[243] Whether it's a technical glitch, a security breach, or a human error, the potential for failure looms over every aspect of organizational functioning. Recognizing this inevitability is the first step towards adopting a proactive and resilient approach to managing setbacks and crises.

Within any company, there are myriad factors that can contribute to failure, ranging from internal complexities to external threats. Operational inefficiencies, inadequate risk management practices, and reliance on outdated technologies are just a few examples of internal factors that can leave companies vulnerable to failure. Additionally, external factors such as market fluctuations, regulatory changes, and geopolitical events can further exacerbate risks and create unforeseen challenges for organizations. By acknowledging the inevitability of failure, companies can cultivate a mindset of preparedness and agility to navigate uncertainties and mitigate potential impacts.

Having a reaction plan in place to deal with problems is paramount for organizations seeking to effectively manage failure when it occurs. A well-defined incident response plan enables companies to respond swiftly and decisively to unexpected disruptions, minimizing their impact on operations and stakeholders.[244] This plan typically includes protocols for detecting, assessing, and containing incidents, as well as procedures for communicating with internal and

---

have an incident response plan." ID Agent at https://www.idagent.com/blog/3-bottom-line-reasons-why-every-business-should-have-an-incident-response-plan/.

[243] PagerDuty at https://business-response.pagerduty.com/what_is/.

[244] *Id.*

external stakeholders. By proactively establishing response mechanisms, companies can streamline decision-making processes and allocate resources more effectively during times of crisis.

A reaction plan serves as a guiding framework for coordinating cross-functional efforts and mobilizing response teams to address the root causes of failure.[245] This may involve deploying technical expertise to resolve technical issues, implementing corrective actions to prevent recurrence, and conducting post-incident reviews to identify lessons learned and areas for improvement.[246] By embracing a culture of continuous learning and improvement, companies can leverage failures as valuable opportunities for growth and innovation.

In the face of failure, organizational leaders play a pivotal role in setting the tone and direction for response efforts. Their ability to demonstrate resilience, decisiveness, and accountability can significantly influence how effectively the company navigates through adversity and emerges stronger on the other side.[247] By fostering a culture of transparency, trust, and collaboration, leaders can empower employees at all levels to contribute their expertise and insights towards overcoming challenges and driving positive change.

Ultimately, the inevitability of failure underscores the importance of proactive risk management and preparedness in today's volatile and uncertain business environment. By

---

[245] Hewitt, Katie (2022). "What is incident Management? Understanding this critical Business Function." Incident.IO at https://incident.io/blog/incident-management.
[246] ID Agent, *supra.*
[247] *Id.*

acknowledging and embracing this reality, companies can cultivate a culture of resilience, adaptability, and innovation that enables them to thrive in the face of adversity and emerge stronger from setbacks.[248] Through effective planning, communication, and leadership, organizations can transform failure from a liability into an opportunity for growth and transformation.

Responding Quickly

In the realm of cybersecurity and incident management, time is of the essence when responding to failures or security breaches. Swift and decisive action is critical to containing the threat, minimizing potential damage, and restoring normal operations as quickly as possible.[249] The longer a security incident remains undetected or unaddressed, the greater the risk of escalating consequences, including data loss, financial losses, and reputational damage.

One of the primary reasons for the importance of rapid response is the concept of "dwell time," which refers to the duration between when an attacker gains unauthorized access to a system and when they are detected and removed. The shorter the dwell time, the less opportunity attackers have to move laterally within the network, exfiltrate sensitive data, or cause further harm.[250] By responding promptly to security alerts and anomalies, organizations can effectively disrupt malicious activities and limit their impact on critical assets and infrastructure.

---

[248] Webb, Rebecca (2021). "4 Reason why every business should draft incident reports." Clear Risk at https://www.clearrisk.com/risk-management-blog/why-you-should-draft-incident-reports.
[249] ID Agent, *supra.*
[250] Hewitt, *supra.*

Timely response is essential for preserving the integrity of digital evidence and conducting thorough forensic investigations into the root causes of security incidents.[251] Rapid containment measures, such as isolating affected systems, disabling compromised accounts, and implementing access controls, can help prevent further compromise and facilitate the collection of forensic data for analysis.[252] This forensic evidence is invaluable for identifying attack vectors, understanding attacker tactics, and strengthening defenses against future threats.

Additionally, prompt response demonstrates organizational readiness and resilience in the face of adversity, instilling confidence among stakeholders, customers, and regulatory authorities.[253] By demonstrating a proactive approach to incident management and adherence to best practices for cybersecurity and data protection, companies can mitigate reputational damage and maintain trust in their ability to safeguard sensitive information. Timely communication with affected parties and stakeholders is also essential for managing public perception and addressing concerns about the security posture of the organization.

In summary, responding quickly to failures or security breaches is essential for containing threats, minimizing damage, and restoring trust and confidence in the organization's ability to protect critical assets and data. By adopting a proactive and agile approach to incident response, organizations can effectively mitigate risks, limit exposure to potential liabilities, and demonstrate their

---

[251] *Id.*
[252] Web, *supra.*
[253] PagerDuty, *supra.*

commitment to safeguarding the interests of stakeholders and customers.[254]

Removing Fear, Employees and their Mistakes

Instilling a culture of fear within an organization, where employees feel apprehensive about admitting mistakes or reporting security incidents due to the fear of punitive measures, can have detrimental effects on cybersecurity and incident response efforts. When employees perceive that making a mistake will result in severe consequences such as reprimand, termination, or loss of reputation, they may be inclined to conceal errors or security breaches rather than addressing them promptly and transparently.[255] This fear-driven behavior can exacerbate the impact of security incidents, hinder effective incident response, and undermine the organization's overall security posture.

One of the primary consequences of a culture of fear is the suppression of incident reporting and communication channels, as employees may hesitate to acknowledge or disclose security incidents for fear of repercussions. This lack of transparency and openness impedes the organization's ability to detect and respond to security threats in a timely manner, allowing issues to escalate and potentially causing greater harm to the organization's assets and reputation.[256] Without accurate and timely reporting of security incidents, decision-makers lack the necessary information to assess risks, allocate resources effectively, and implement appropriate mitigation measures.

Next, a culture of fear can erode trust and collaboration within the organization, creating barriers to effective

---

[254] Web, *supra.*
[255] Hewitt, *supra.*
[256] *Id.*

teamwork and information sharing among employees and departments. When individuals are more concerned about protecting themselves from blame or punishment than working together to address security challenges collectively, it can lead to siloed behavior, communication breakdowns, and missed opportunities for proactive risk management. In such environments, employees may be reluctant to seek assistance or guidance from colleagues or security experts, further exacerbating vulnerabilities and increasing the likelihood of security incidents going unnoticed or unresolved.

A culture of fear can stifle innovation and creativity within the organization, as employees may be hesitant to experiment or propose new ideas for fear of failure or reprisal.[257] In dynamic and rapidly evolving cybersecurity landscapes, where adaptive and innovative approaches are essential for staying ahead of emerging threats, fostering a culture of risk aversion can impede the organization's ability to adapt, innovate, and effectively respond to evolving security challenges. By discouraging experimentation and risk-taking, organizations may miss out on valuable opportunities to identify and address vulnerabilities before they are exploited by malicious actors.

Ultimately, organizations must strive to cultivate a culture of trust, transparency, and accountability, where employees feel empowered to report security incidents, admit mistakes, and collaborate openly to address security challenges. By promoting a supportive and non-punitive environment that encourages learning from failures, sharing knowledge, and embracing constructive feedback,

---

[257] Web, *supra*.

organizations can strengthen their cybersecurity resilience, foster a culture of continuous improvement, and empower employees to play an active role in safeguarding the organization's assets and reputation.

Public Relations

Having a well-prepared public relations (PR) plan is crucial for managing the aftermath of a security breach effectively and mitigating potential reputational damage to the organization.[258] In today's interconnected and information-savvy world, security incidents can quickly attract media attention, public scrutiny, and stakeholder concern, making it essential for organizations to respond promptly, transparently, and effectively to protect their reputation and maintain stakeholder trust.

First and foremost, a robust PR plan enables organizations to communicate proactively with key stakeholders, including customers, employees, investors, regulators, and the general public, providing timely updates, accurate information, and reassurance about the steps being taken to address the security breach and mitigate its impact.[259] By demonstrating transparency, accountability, and a commitment to resolving the issue, organizations can foster trust and confidence among stakeholders, helping to minimize reputational harm and preserve goodwill.

Moreover, a well-executed PR plan helps organizations control the narrative surrounding the security breach, shaping public perception and framing the incident in a way that emphasizes the organization's response efforts, commitment to cybersecurity, and dedication to protecting

---

[258] Hewitt, *supra*.
[259] *Id.*

stakeholder interests.[260] Through strategic messaging and media engagement, organizations can influence how the incident is portrayed in the media, mitigate negative coverage, and position themselves as proactive and responsible stewards of data security and privacy.

Additionally, a comprehensive PR plan includes provisions for managing social media channels and online communication platforms, enabling organizations to monitor and respond to public sentiment, address inquiries and concerns from stakeholders, and counter misinformation or rumors in real-time.[261] By maintaining an active and transparent online presence during a security breach, organizations can demonstrate their commitment to openness, accountability, and stakeholder engagement, helping to build and preserve trust in the face of adversity.

Ultimately, an effective PR plan serves as a critical component of crisis management and reputation protection, enabling organizations to navigate the complexities of a security breach, mitigate reputational risks, and emerge from the incident with their reputation intact. By prioritizing transparency, communication, and stakeholder engagement, organizations can strengthen their resilience in the face of cybersecurity threats and demonstrate their unwavering commitment to safeguarding the interests of their stakeholders.

Legal Action Against Attacker

Fast and hostile legal action against hackers or hacker groups can serve as a powerful deterrent against future attacks by sending a clear message that cybercriminal

---

[260] Web, *supra*.
[261] *Id.*

activities will not be tolerated and will be met with swift and severe consequences.[262] By taking aggressive legal action, organizations demonstrate their commitment to defending their interests, protecting their assets, and holding perpetrators accountable for their actions, thereby dissuading potential attackers and deterring similar incidents in the future.

Aggressive legal action can disrupt the operations of cybercriminals and dismantle their networks, making it more difficult for them to carry out further attacks and reducing their ability to cause harm to other organizations and individuals. By targeting the perpetrators behind the security breach and seeking legal recourse through civil lawsuits, criminal charges, or law enforcement cooperation, organizations can disrupt the cybercriminal ecosystem, disrupt their activities, and undermine their capabilities, thereby enhancing overall cybersecurity posture and resilience.[263]

Fast and hostile legal action can serve as a form of retaliation and retribution against cybercriminals, sending a clear message that their actions will not go unpunished and that they will be held accountable for the damage they cause.[264] By aggressively pursuing legal remedies, organizations can seek restitution for damages, recover stolen assets, and seek punitive damages to deter future misconduct and compensate for the harm inflicted on the organization, its stakeholders, and affected individuals.

Additionally, aggressive legal action can help organizations assert their rights, protect their intellectual property, and

---

[262] PagerDuty, *surpa.*
[263] *Id.*
[264] Hewitt, *supra.*

defend their proprietary information from unauthorized access, theft, or exploitation.[265] By vigorously defending their legal interests and seeking legal redress for any violations of their rights or breaches of their security, organizations can safeguard their competitive advantage, preserve their reputation, and maintain trust and confidence among their customers, partners, and stakeholders.

Fast and hostile legal action against hackers or hacker groups can be an effective strategy for mitigating future attacks by deterring potential perpetrators, disrupting cybercriminal activities, seeking retribution and restitution, asserting legal rights, and enhancing overall cybersecurity resilience.[266] By leveraging legal remedies and law enforcement cooperation to hold cybercriminals accountable for their actions, organizations can send a strong message that cybersecurity breaches will not be tolerated and that those responsible will face severe consequences for their misconduct.

Conclusion

Chapter 8 underscores the critical importance of effective incident response strategies and public relations efforts in managing major security breaches within web3 companies. The inevitability of failure in any organization highlights the necessity of having robust response plans in place to address unforeseen challenges swiftly and effectively. By acknowledging the inevitability of failure and preparing comprehensive incident response protocols, organizations can minimize the impact of security incidents and mitigate potential damage to their operations, reputation, and stakeholders.

---

[265] *Id.*

[266] PagerDuty, *supra.*

Responding quickly to security breaches is paramount to containing the threat and preventing further escalation of the situation. Rapid response allows organizations to identify and isolate compromised systems, mitigate the spread of malicious activity, and implement remediation measures to restore normal operations. Timely intervention helps limit the extent of damage and facilitates the recovery process, enabling organizations to resume business activities with minimal disruption and downtime.

Fostering a culture of transparency and trust within the organization is essential to encourage employees to report incidents promptly without fear of reprisal. Creating an environment where individuals feel empowered to acknowledge mistakes and seek assistance fosters a proactive approach to cybersecurity and enables organizations to address vulnerabilities and threats more effectively. By removing fear and promoting open communication, organizations can strengthen their security posture and enhance their ability to detect, respond to, and mitigate security incidents.

Effective public relations strategies play a crucial role in managing the aftermath of security breaches and preserving organizational reputation and stakeholder trust. Transparent communication, timely updates, and honest disclosure of information help maintain credibility and demonstrate accountability, fostering trust and confidence among customers, partners, and the public. By proactively managing the narrative surrounding security incidents and demonstrating a commitment to cybersecurity and customer protection, organizations can mitigate reputational damage and preserve brand integrity in the face of adversity.

Chapter 9: Insider Threats

"If you don't feel ordained by the Universe to do this job, do something else. The intelligence community has to shut down the gaping wound that is the insider threat epidemic we are experiencing right now."

Jamse Scott[267]

Chapter 9 delves into the critical yet often overlooked aspect of insider threats within web3 companies, highlighting the imperative for organizations to vigilantly assess their hires and mitigate the risks posed by potentially malicious insiders. As organizations entrust employees with access to sensitive data, systems, and resources, the threat of insider threats looms large, underscoring the need for robust vetting processes and proactive security measures. By recognizing the potential for insider threats and implementing stringent safeguards, organizations can fortify their defenses and safeguard against internal risks.

The notion of "Too Good to Be True" serves as a cautionary reminder of the potential dangers associated with employees who exhibit exceptional qualifications, credentials, or skills. While skilled and talented individuals can undoubtedly contribute to organizational success, organizations must exercise due diligence in scrutinizing candidates who may appear overly proficient or well-qualified. In some cases, individuals with exaggerated credentials or deceptive intentions may infiltrate organizations with the aim of exploiting their access privileges for nefarious purposes. By maintaining a healthy skepticism and conducting thorough background checks,

[267] (2024). "Insider Threats Quotes, Jamse Scott." Good Reads at https://www.goodreads.com/quotes/tag/insider-threat.

organizations can identify red flags and mitigate the risks associated with deceptive hires.

Background checks play a pivotal role in the hiring process, enabling organizations to verify the credentials, employment history, and criminal records of prospective employees. By conducting comprehensive background checks, organizations can validate the authenticity of candidates' claims and uncover any discrepancies or concerning information that may indicate a propensity for insider threats. Moreover, background checks empower organizations to make informed hiring decisions and mitigate the risks of onboarding individuals with a history of misconduct or criminal behavior.

Disgruntled employees pose a significant insider threat to organizational security, as their grievances or grievances may motivate them to engage in malicious activities aimed at sabotaging or harming the organization. Whether motivated by personal grievances, financial incentives, or ideological reasons, disgruntled employees can leverage their insider access to perpetrate data breaches, sabotage systems, or engage in other malicious activities. To mitigate the risks posed by disgruntled employees, organizations must foster a positive work culture, address employee grievances promptly and constructively, and implement robust security measures to monitor and detect anomalous behavior.

The threat of insider threats underscores the importance of implementing robust vetting processes, conducting thorough background checks, and proactively monitoring employee behavior to mitigate the risks posed by potentially malicious insiders. By remaining vigilant and proactive in identifying and addressing insider threats, organizations can safeguard their data, assets, and reputation from internal risks and uphold the integrity and security of their operations in an increasingly interconnected and digital business landscape.

Too Good to be True

When confronted with a candidate who appears "too good to be true," organizations must exercise caution and diligence to ensure they aren't unwittingly welcoming a potential security risk into their midst.[268] The first step in addressing such a situation involves conducting a thorough assessment of the candidate's qualifications, experience, and credentials. While impressive resumes and exceptional skills may be genuine, they can also serve as red flags warranting closer scrutiny. Organizations should verify the authenticity of the candidate's claims through rigorous background checks, reference checks, and credential verification procedures.[269]

Furthermore, it's essential for organizations to delve deeper into the candidate's professional background and track record to ascertain the veracity of their claims and identify any inconsistencies or discrepancies. This may involve scrutinizing the candidate's employment history, educational credentials, certifications, and any past accomplishments or accolades.[270] By corroborating the information provided by the candidate and cross-referencing it with independent sources, organizations can

---

[268] Ahonen, Elias (2022). "Fake Employees and Social Attacks: Crypto Recruiting is a minefield." Magazine by CoinTelegraph at https://cointelegraph.com/magazine/fake-employees-social-attacks-crypto-recruiting-minefield/.
[269] *Id.*
[270] Beller, Morgan (2022). "Why Web3 Hiring is Unique." NFX at https://www.nfx.com/post/why-web3-hiring-is-unique.

gain a more accurate understanding of the candidate's qualifications and suitability for the role.

In addition to verifying credentials, organizations should also assess the candidate's cultural fit, interpersonal skills, and alignment with the organization's values and mission. While technical expertise is undoubtedly important, it's equally crucial for candidates to demonstrate integrity, professionalism, and ethical conduct. Hiring decisions should not be based solely on technical prowess but should also consider factors such as trustworthiness, reliability, and compatibility with the organization's culture.

Organizations should be mindful of potential warning signs or red flags that may indicate the candidate's intentions or character. These could include inconsistencies in the candidate's narrative, evasive responses to questions, reluctance to provide references, or gaps in employment history.[271] While not conclusive evidence of malicious intent, such indicators should prompt organizations to proceed with caution and conduct additional due diligence to mitigate potential risks.

Ultimately, when faced with a candidate who appears "too good to be true," organizations must prioritize thorough vetting, due diligence, and risk assessment to safeguard against potential security threats. By adopting a proactive and cautious approach to hiring decisions, organizations can mitigate the risks of inadvertently welcoming security risks into their ranks and ensure the integrity and security of their operations.

---

[271] *Id.*

Background Checks

Background checks play a pivotal role in enhancing the security posture of web3 companies by providing valuable insights into the backgrounds, histories, and potential risks associated with individuals seeking employment or access to sensitive information and systems.[272] In the dynamic and rapidly evolving landscape of web3 technologies, where data breaches and cyber threats loom large, the importance of conducting thorough background checks cannot be overstated. These checks serve as a critical line of defense against insider threats, malicious actors, and other security vulnerabilities that may compromise the integrity and confidentiality of organizational assets.

One of the primary reasons background checks are indispensable for web3 companies is their ability to verify the authenticity of candidates' credentials, qualifications, and employment histories.[273] In an industry characterized by innovation and digital transformation, where technical expertise and specialized skills are highly valued, ensuring the accuracy and legitimacy of applicants' claims is paramount.[274] Background checks enable organizations to validate the educational backgrounds, professional certifications, and work experience of candidates, helping to weed out individuals who may have misrepresented their qualifications or fabricated their credentials.

Background checks enable web3 companies to identify and assess potential red flags or risk factors that may pose

---

[272] MeasureOne at https://www.measureone.com/customers/web3.
[273] Daigle, Zach (2023). "Velocity Network: The Future of Background Screening." Ocisive at https://blog.cisive.com/velocity-network-the-future-of-background-screening.
[274] *Id.*

security threats or reputational risks to the organization.[275] By examining candidates' criminal records, credit histories, and professional reputations, organizations can uncover any past indiscretions, legal entanglements, or unethical behaviors that may raise concerns about their trustworthiness or suitability for employment.[276] By proactively screening candidates for potential security risks, web3 companies can mitigate the likelihood of insider threats, data breaches, and other security incidents, thereby safeguarding their assets, reputation, and stakeholders' interests.

Disgruntled Employees

Dealing with disgruntled employees is a critical aspect of maintaining security within web3 companies, as such individuals can pose significant risks to the organization's data, systems, and operations.[277] By implementing proactive measures to address and mitigate the factors contributing to employee dissatisfaction, organizations can reduce the likelihood of disgruntled employees becoming security risks. Here are five strategies for effectively managing and mitigating this threat:

Firstly, fostering a positive work environment characterized by open communication, transparency, and mutual respect can help prevent employees from becoming disgruntled in the first place.[278] By promoting a culture of inclusivity,

---

275 MeasureOne, *supra*.
276 Staff (2023). Web3 Identity: Why you cant get a crypto job." Netizens at https://medium.com/@NetizensNFT/web3-identity-why-you-cant-get-a-crypto-job-bad521f54efd.
277 Zacamos (2023). "One of the Top Cybersecurity Risks in your business is employee apathy." Hackernoon at https://hackernoon.com/one-of-the-top-cybersecurity-risks-in-your-business-is-employee-apathy.

recognition, and support, organizations can enhance employee satisfaction and morale, reducing the likelihood of grievances and grievances escalating into security concerns.

Secondly, implementing robust conflict resolution mechanisms and channels for employees to voice their concerns and grievances in a constructive manner can help address underlying issues before they escalate.[279] Providing avenues for employees to seek assistance, raise complaints, and resolve conflicts internally can prevent grievances from festering and potentially leading to security breaches or malicious behavior.

Thirdly, conducting regular employee satisfaction surveys, performance evaluations, and feedback sessions can help identify and address sources of dissatisfaction or discontent among employees.[280] By proactively soliciting feedback and addressing concerns in a timely and transparent manner, organizations can demonstrate their commitment to addressing employee needs and improving the overall work environment.

Fourthly, establishing clear policies, procedures, and protocols for addressing employee grievances, disputes, and disciplinary actions can help ensure consistency, fairness, and accountability in dealing with internal conflicts and misconduct.[281] By clearly outlining

---

[278] *Id.*
[279] Dioquino, Vince (2024). "eBay lays of 30% of Web3 and NFT staff-report." Crypto Briefing at https://cryptobriefing.com/ebay-lays-off-30-of-web3-and-nft-staff/.
[280] Zacamos, *supra.*
[281] Canales, Katie (2022). The Crypto CEO who made waves for paying disgruntled employees to leave says he has no regrets about offending anyone." Business Insider at

expectations, consequences, and avenues for recourse, organizations can deter potential security risks and promote a culture of compliance and accountability among employees.

Lastly, providing training and education on cybersecurity best practices, data protection policies, and ethical conduct can help raise awareness among employees about their responsibilities and obligations in safeguarding organizational assets and information.[282] By empowering employees with the knowledge and skills to recognize and report security threats, organizations can enlist their support in maintaining a vigilant and proactive stance against insider risks and malicious activities.

Conclusion

In conclusion, mitigating insider threats is paramount for ensuring the security and integrity of web3 companies' systems, data, and operations. By recognizing the potential risks posed by disgruntled employees and implementing proactive measures to address them, organizations can effectively safeguard against internal vulnerabilities and protect against malicious activities. Through strategies such as fostering a positive work environment, implementing robust conflict resolution mechanisms, conducting regular employee feedback sessions, and establishing clear policies and procedures, companies can create a culture of accountability, transparency, and trust that deters insider

---

https://www.businessinsider.in/cryptocurrency/news/the-crypto-ceo-who-made-waves-for-paying-disgruntled-employees-to-leave-says-he-has-no-regrets-about-offending-anyone/articleshow/92709930.cms.

[282] Staff (2024). "Employer v. Contractor: Misclassification risk in crypto." TOKU at https://www.toku.com/article/employer-vs-contractor-misclassification-risk-in-crypto.

threats and promotes a sense of collective responsibility for security.

The importance of conducting thorough background checks during the hiring process cannot be overstated, as it allows organizations to identify and mitigate potential risks early on. By vetting candidates carefully and assessing their suitability for positions that involve access to sensitive information or critical systems, companies can reduce the likelihood of insider threats emerging from within their ranks. Additionally, providing ongoing training and education on cybersecurity best practices and ethical conduct can empower employees to recognize and report suspicious behavior, further bolstering the organization's defenses against insider risks.

Addressing insider threats requires a multifaceted approach that encompasses proactive measures to prevent dissatisfaction and misconduct, as well as reactive strategies to detect and respond to security incidents swiftly and effectively. By prioritizing the establishment of a positive work culture, investing in employee development and education, and implementing robust security protocols and procedures, web3 companies can fortify their defenses against insider threats and uphold the trust and confidence of their stakeholders in an increasingly digitized and interconnected world.

Conclusion

In conclusion, Part II of this book has delved into the intricate landscape of account security within the realm of web3 technologies, highlighting the multifaceted challenges and proactive measures necessary to safeguard against various threats and vulnerabilities. From password management and encryption protocols to incident response strategies and insider threat mitigation, each chapter has provided valuable insights and practical recommendations for enhancing the security posture of web3 companies.

The chapters have underscored the critical importance of adopting robust password policies, leveraging encryption technologies, and staying vigilant against evolving cyber threats. With the proliferation of sophisticated attacks and the increasing interconnectedness of digital ecosystems, organizations must remain proactive in fortifying their defenses and mitigating risks to protect sensitive data, intellectual property, and customer trust.

Furthermore, the discussions on incident response planning, public relations management, and legal considerations have emphasized the importance of preparedness and resilience in the face of security breaches. By implementing comprehensive response strategies and fostering a culture of transparency and accountability, companies can effectively navigate crises and mitigate potential reputational and financial damages.

Additionally, the chapters have highlighted the significance of addressing insider threats and ensuring thorough vetting processes during the hiring phase. By prioritizing employee education, implementing stringent access controls, and fostering a positive work culture, organizations can minimize the risk of internal breaches and unauthorized access to critical assets.

Looking ahead, it is clear that the landscape of web3 security will continue to evolve alongside technological advancements and emerging threats. As companies embrace decentralization, blockchain technologies, and the Internet of Things (IoT), they must remain adaptive and proactive in implementing robust security measures that align with the principles of accountability, transparency, and trust.

In essence, Part II has provided a comprehensive framework for enhancing account security in the web3 era, emphasizing the importance of proactive risk management, continuous improvement, and collaboration across stakeholders. By embracing these principles and leveraging the insights shared in this book, organizations can strengthen their resilience to cyber threats and uphold the integrity and trust of their digital ecosystems in an increasingly complex and interconnected world.

Part III: Social Engineering and Phishing

Introduction Part III

Part III of this book delves deep into the critical realm of social engineering and phishing, shedding light on the challenges posed by these pervasive cyber threats and offering actionable insights to fortify organizational defenses. In today's digital landscape, where sophisticated cyberattacks continue to proliferate and threaten the integrity of businesses worldwide, cultivating a culture of security is paramount to mitigate risks, protect sensitive information, and safeguard organizational assets.

As organizations grapple with the ever-evolving tactics employed by malicious actors to exploit human vulnerabilities and gain unauthorized access to sensitive data, understanding the nuances of social engineering and phishing becomes essential. Part III serves as a comprehensive guide to navigating these complex threats, equipping readers with the knowledge, strategies, and best practices necessary to fortify their defenses and defend against potential breaches.

The chapters in this section are meticulously crafted to address key facets of social engineering and phishing, from the psychology behind these attacks to the practical steps organizations can take to mitigate risks and enhance their security posture. By examining real-world scenarios, case studies, and industry insights, Part III offers a holistic perspective on the challenges and opportunities associated with combating social engineering and phishing in today's digital age.

At the heart of Part III lies the recognition that social engineering and phishing attacks often exploit human psychology and trust to manipulate individuals into divulging sensitive information or performing malicious actions. By understanding the tactics employed by attackers and the behavioral cues that may indicate a potential threat, organizations can empower their employees to recognize and resist social engineering attempts, thereby reducing the likelihood of successful attacks.

Part III underscores the importance of proactive measures, such as security awareness training, simulated phishing exercises, and robust access controls, in building a resilient defense against social engineering and phishing attacks. By investing in education, awareness, and technological solutions, organizations can strengthen their security posture and mitigate the risks posed by these insidious threats.

The chapters in this section delve into the intricacies of security protocols, encryption methods, and incident response strategies tailored to combat social engineering and phishing attacks. From implementing multi-factor authentication and encryption measures to developing robust incident response plans and security policies, organizations can enhance their readiness to detect, respond to, and recover from potential breaches.

In summary, Part III serves as a comprehensive resource for organizations seeking to navigate the complex landscape of social engineering and phishing threats. By fostering a culture of security, leveraging technology and best practices, and staying vigilant against emerging threats, organizations can fortify their defenses and protect their most valuable assets in an increasingly digital and interconnected world.

Chapter 10: Multi-Factor Identification

"Turn on all security features like two-factor authentication. People who do that generally don't get hacked. Don't care? You will when you get hacked. Do the same for your email and other social services too."

Robert Scoble[283]

In this chapter, we delve into the critical realm of multi-factor identification (MFA), an essential component of modern cybersecurity strategies aimed at enhancing access control and thwarting unauthorized access attempts. By requiring users to provide multiple forms of verification before granting access to sensitive systems or data, MFA significantly strengthens authentication mechanisms and reduces the risk of unauthorized breaches.[284] This chapter explores various aspects of MFA implementation, including notable companies with robust 2FA practices, the benefits of using multiple devices for authentication, and the importance of regularly upgrading systems to maintain security efficacy.

Within the chapter, we examine exemplary companies that have implemented robust two-factor authentication (2FA) solutions to enhance security and protect sensitive information. By requiring users to provide two or more forms of authentication, such as a password and a one-time code sent to their mobile device, these organizations significantly reduce the risk of unauthorized access and

---

[283] (2024). "Multi-Factor Authentication, Robert Scoble." Brainy Quote at https://www.brainyquote.com/quotes/robert_scoble_707410.
[284] Staff (2022). "Multi-Factor Authentication" CISA at https://www.cisa.gov/resources-tools/resources/multi-factor-authentication-mfa#:~:text=Multifactor%20authentication%20is%20a%20layered,a%20user%27s%20identity%20for%20login..

strengthen their overall security posture. Examples of companies known for their strong 2FA practices include leading tech giants, financial institutions, and government agencies that prioritize security and invest in advanced authentication technologies.

Further, we discuss the benefits of leveraging multiple devices for authentication purposes, such as using a combination of smartphones, tokens, or biometric devices to verify users' identities. By diversifying authentication factors across different devices, organizations can enhance security resilience and mitigate the risk of single-point failures or device compromises. Additionally, multi-device authentication provides users with greater flexibility and convenience while maintaining robust security standards, ensuring a seamless and secure user experience across various platforms and environments.

This chapter underscores the importance of regularly upgrading systems and authentication mechanisms to adapt to evolving security threats and maintain optimal protection against unauthorized access attempts. Outdated systems and authentication methods may be vulnerable to exploitation by cybercriminals, highlighting the critical need for organizations to stay abreast of emerging threats and implement timely updates and patches. By proactively upgrading systems, software, and authentication protocols, organizations can mitigate security vulnerabilities, enhance resilience against cyber threats, and safeguard their valuable assets and sensitive information.

In summary, this chapter explores the fundamental principles of multi-factor identification (MFA) and its pivotal role in modern cybersecurity practices. By examining exemplary companies with robust 2FA practices, highlighting the benefits of leveraging multiple devices for authentication, and emphasizing the importance of regular system upgrades, organizations can strengthen their access control measures, mitigate the risk of unauthorized breaches, and uphold the integrity and confidentiality of their digital assets.

Good Companies, Good Policies

A good multi-factor identification (MFA) company is characterized by several key attributes that collectively contribute to robust security measures and user-friendly experiences.[285] Firstly, such companies prioritize the implementation of multiple authentication factors beyond traditional passwords, enhancing security by requiring users to verify their identity through a combination of different methods.[286] These factors typically include something the user knows (e.g., passwords), something they have (e.g., a mobile device or security token), and something they are (e.g., biometric data like fingerprints or facial recognition), thereby significantly reducing the likelihood of unauthorized access through credential theft or brute-force attacks.

Secondly, effective MFA companies leverage advanced technologies and encryption protocols to safeguard user credentials and sensitive information throughout the authentication process.[287] By employing robust encryption algorithms and secure communication channels, these companies ensure that user data remains protected from interception or unauthorized access, bolstering overall security and compliance with regulatory requirements.

---

285 *Id.*
286 Yarar, Kinza (2024). "Multifactor Authentication." TechTarget at https://www.techtarget.com/searchsecurity/definition/multifactor-authentication-MFA.
287 CISA, *supra.*

Additionally, they regularly update their security infrastructure and software systems to address emerging threats and vulnerabilities, staying ahead of potential attackers and mitigating security risks proactively.

A good MFA company prioritizes user convenience and seamless integration with existing workflows and applications. By offering intuitive and user-friendly authentication methods, such as push notifications, biometric authentication, or one-time passcodes, they streamline the user experience while maintaining strong security measures.[288] This balance between security and usability is essential for encouraging widespread adoption of MFA solutions across diverse user populations and organizational settings, ensuring broad protection against unauthorized access without imposing undue complexity or inconvenience on users.

Reputable MFA companies demonstrate a commitment to continuous innovation and adaptation to evolving security threats and industry standards. They invest in research and development efforts to explore emerging technologies, such as artificial intelligence, machine learning, or behavioral analytics, that can enhance authentication processes and threat detection capabilities.[289] By staying at the forefront of technological advancements and industry best practices, these companies remain well-equipped to address new challenges and evolving attack vectors, empowering their clients with cutting-edge security solutions tailored to their unique needs and risk profiles.

---

[288] Staff (2024). "What is Multi-Factor Authentication (MFA) and How Does it Work?" One Login at https://www.onelogin.com/learn/what-is-mfa.
[289] Amazon Web Services at https://aws.amazon.com/what-is/mfa/.

Multiple Devices

Having multiple devices in a two-factor authentication (2FA) system is crucial for enhancing security and ensuring reliable access controls.[290] Firstly, the use of multiple devices adds an extra layer of redundancy to the authentication process, reducing the risk of being locked out of an account due to device loss, damage, or technical issues. By providing users with alternative authentication options across different devices (e.g., smartphones, tablets, or hardware tokens), organizations minimize the impact of device-related disruptions and ensure continuous access to critical systems and resources.

Secondly, the presence of multiple devices enables organizations to implement diverse authentication factors and distribution channels, thereby strengthening overall security posture.[291] By offering users the flexibility to choose between different authentication methods and devices (e.g., biometric authentication on a smartphone or hardware token authentication via USB), organizations can tailor their 2FA solutions to suit varying user preferences, risk profiles, and operational requirements.[292] This diversity in authentication factors enhances security by mitigating the risk of single-point failures or vulnerabilities associated with specific devices or methods, fostering resilience against sophisticated attacks and social engineering tactics.

Next, the use of multiple devices in 2FA systems enhances user convenience and accessibility, accommodating diverse user populations with varying technological capabilities and preferences.[293] By supporting a wide range of devices

---

[290] Gibson, *infra*.
[291] CISA, *supra*.
[292] *Id.*
[293] AWS, *supra*.

and platforms, organizations empower users to leverage familiar and convenient authentication methods that align with their workflow and device preferences. This inclusivity promotes user adoption and compliance with security policies, as users are more likely to embrace 2FA measures that integrate seamlessly into their existing habits and devices, without requiring significant changes to their routines or behaviors.

The incorporation of multiple devices in 2FA systems enhances security through device separation and isolation, reducing the likelihood of unauthorized access or compromise. By distributing authentication factors across separate devices (e.g., combining a password entered on a computer with a one-time passcode generated on a smartphone), organizations create distinct authentication channels that increase the complexity of attacks and mitigate the risk of account compromise through single-vector attacks.[294] This compartmentalization of authentication factors enhances security by reducing the attack surface and isolating potential vulnerabilities, making it harder for adversaries to gain unauthorized access to sensitive accounts or data.

Lastly, the use of multiple devices in 2FA systems facilitates scalability and flexibility, allowing organizations to adapt their authentication mechanisms to evolving security requirements and technological advancements.[295] By supporting a diverse array of devices and authentication methods, organizations can future-proof their 2FA solutions and accommodate changes in user preferences, device capabilities, or regulatory mandates over time. This

---

[294] *Id.*
[295] Gibson, Jessica (2022). "Greater Digital Risks Loom Over People with Multiple Devices." Cyber at https://cybermagazine.com/cyber-security/greater-digital-risks-loom-over-people-with-multiple-devices.

adaptability enables organizations to maintain robust security measures while remaining agile and responsive to shifting threat landscapes and business needs, ensuring long-term resilience and effectiveness in defending against unauthorized access and cyber threats.[296]

Upgrading Systems

Constantly upgrading systems is essential for maintaining robust security measures and mitigating evolving cyber threats effectively. Firstly, regular system upgrades help address known vulnerabilities and security weaknesses identified through ongoing risk assessments, security audits, and threat intelligence analysis.[297] By promptly applying security patches, software updates, and firmware upgrades, organizations can remediate known vulnerabilities before they can be exploited by malicious actors, reducing the risk of successful cyberattacks and data breaches. This proactive approach to system maintenance minimizes exposure to known vulnerabilities and enhances overall resilience against common attack vectors, such as malware infections, remote code execution, or privilege escalation exploits.[298]

Secondly, system upgrades enable organizations to leverage the latest security technologies, features, and best practices to enhance their defensive capabilities and stay ahead of emerging threats.[299] As cyber threats continue to

---

[296] Staff (2024). "Understanding Web3 and its security implications." The Net at https://www.cloudflare.com/the-net/web3-security/.
[297] CISA, *supra*.
[298] Hooper, Anatol (2023). "Web3 protection tool releases update that improves security against scam tokens and dangerous contracts." Coin Telegraph at https://cointelegraph.com/news/web3-protection-tool-releases-update-that-improves-security-against-scam-tokens-and-dangerous-contracts.

evolve in sophistication and complexity, outdated or legacy systems may lack the necessary security controls and countermeasures to effectively protect against modern attack methods and exploitation techniques. By upgrading systems to newer versions or adopting cutting-edge security solutions, organizations can access advanced security features, such as enhanced encryption algorithms, intrusion detection systems, behavioral analytics, or threat intelligence integration, to bolster their cyber defense posture and adapt to changing threat landscapes.

Regular system upgrades contribute to regulatory compliance and adherence to industry standards by ensuring that organizations maintain up-to-date software versions and security configurations in alignment with legal requirements, industry guidelines, and security best practices.[300] Many regulatory frameworks and compliance mandates mandate regular system maintenance, patch management, and vulnerability remediation as essential components of an organization's cybersecurity program. By demonstrating a commitment to continuous improvement and proactive risk management through timely system upgrades, organizations can mitigate compliance risks, avoid potential penalties or fines, and build trust with customers, partners, and regulatory authorities regarding their commitment to safeguarding sensitive data and maintaining robust security controls.

Conclusion

In conclusion, multi-factor identification (MFA) stands as a critical pillar in modern cybersecurity strategies, providing an additional layer of defense against unauthorized access

---

[299] *Id.*
[300] CISA, *supra.*

and identity theft. Through the implementation of robust MFA solutions, organizations can significantly enhance the security of their digital assets, sensitive information, and critical systems by requiring users to provide multiple forms of verification before granting access. By leveraging a combination of factors such as passwords, biometrics, hardware tokens, or mobile devices, MFA mechanisms strengthen authentication processes and reduce the likelihood of successful credential-based attacks, such as phishing, brute-force attacks, or password spraying.

The effectiveness of MFA hinges on the selection of reliable and trustworthy solutions from reputable vendors with a proven track record of delivering secure authentication technologies. Organizations must prioritize solutions that offer strong encryption, tamper-resistant hardware, and robust authentication protocols to safeguard against interception, tampering, or compromise. By partnering with reputable MFA providers and adhering to industry best practices, organizations can deploy MFA solutions that meet their specific security requirements while minimizing the risk of vulnerabilities or weaknesses that could undermine their effectiveness.

Additionally, the adoption of MFA should extend beyond traditional user authentication scenarios to encompass various use cases, including remote access, privileged account management, and application-specific authentication requirements. By implementing MFA across diverse IT environments and access scenarios, organizations can establish a comprehensive defense-in-depth strategy that mitigates the risk of unauthorized access and strengthens overall security posture. Furthermore, organizations should prioritize user education and awareness initiatives to promote the importance of MFA adoption, encourage compliance with security policies, and empower users to recognize and respond appropriately to potential security threats or phishing attempts targeting their credentials.

In summary, the deployment of multi-factor identification represents a crucial component of modern cybersecurity strategies, offering organizations a powerful means of fortifying their defenses against evolving threats and safeguarding sensitive information from unauthorized access or misuse. By embracing MFA solutions that prioritize security, usability, and scalability, organizations can enhance authentication processes, reduce the risk of account compromise, and foster a culture of security-conscious behavior among users, ultimately strengthening the resilience of their digital infrastructure in the face of emerging cyber threats.

Chapter 11: Phishing Attacks

"A Nation State or Cyber-Mercenary won't hack e-voting machines one by one. This takes too long and will have minimal impact. Instead, they'll take an easier approach like spear phishing the manufacturer with malware and poison the voting machine update pre-election and allow the manufacturer to update each individual machine with a self-deleting payload that will target the tabulation process."

James Scott[301]

Phishing is a form of cyberattack wherein malicious actors employ deceptive tactics, such as fraudulent emails, text messages, or websites, to trick individuals into divulging sensitive information, such as login credentials, financial details, or personal data. Typically, phishing attempts masquerade as legitimate communications from reputable entities, such as banks, government agencies, or trusted organizations, aiming to deceive recipients into unwittingly disclosing confidential information or performing actions that compromise their security.[302] By exploiting human vulnerabilities and leveraging social engineering techniques, phishing attacks manipulate users' trust and curiosity to gain unauthorized access to sensitive information or facilitate fraudulent activities, posing significant risks to individuals, businesses, and organizations worldwide.

---

[301] (2024). "Phishing Quotes, James Scott." Good Reads at https://www.goodreads.com/quotes/tag/phishing.
[302] Staff (2023). "State of the Phish Report" Proofpoint at https://www.proofpoint.com/us/threat-reference/phishing.

Chapter 11 delves into the intricate world of phishing attacks, a prevalent form of cyber threat that exploits human vulnerabilities to deceive individuals and organizations into divulging sensitive information or performing malicious actions. Phishing attacks typically involve the use of deceptive techniques, such as fraudulent emails, fake websites, or social engineering tactics, to trick unsuspecting victims into disclosing personal credentials, financial details, or other confidential data. By masquerading as legitimate entities or trusted sources, malicious actors seek to exploit human trust and curiosity to achieve their nefarious objectives, whether it be stealing sensitive information, spreading malware, or gaining unauthorized access to computer systems.

Common tactics employed in phishing attacks encompass a wide range of strategies designed to manipulate human psychology and exploit weaknesses in cybersecurity defenses. These tactics may include the creation of convincing email messages that mimic authentic communications from reputable organizations, the use of urgency or fear-inducing language to prompt immediate action, the impersonation of trusted individuals or authorities to establish credibility, and the insertion of malicious links or attachments to compromise victims' devices or accounts. Additionally, phishing attacks often leverage social engineering techniques to manipulate human behavior, such as preying on emotions, exploiting curiosity, or capitalizing on trust relationships to deceive victims and elicit desired responses.

To mitigate the risk of falling victim to phishing attacks, individuals and organizations must adopt proactive measures to enhance their resilience against this pervasive threat. Key strategies for avoiding phishing attacks include maintaining a healthy skepticism towards unsolicited emails, messages, or requests, scrutinizing sender information and email content for signs of suspicious activity or inconsistencies, avoiding clicking on links or downloading attachments from unknown or untrusted sources, and verifying the legitimacy of requests for sensitive information through alternative channels or direct communication with known contacts. Furthermore, organizations should invest in comprehensive cybersecurity awareness training programs to educate employees about the dangers of phishing attacks, promote best practices for identifying and reporting suspicious activities, and cultivate a culture of vigilance and security consciousness across the workforce.

Phishing attacks represent a significant cybersecurity challenge that continues to pose a threat to individuals, businesses, and organizations worldwide. By understanding the nature of phishing attacks, recognizing common tactics employed by malicious actors, and implementing effective countermeasures to prevent and mitigate their impact, individuals and organizations can bolster their defenses against this insidious threat and safeguard their sensitive information, financial assets, and digital infrastructure from exploitation and compromise. Through a combination of awareness, vigilance, and proactive security measures, stakeholders can reduce their susceptibility to phishing attacks and contribute to the collective effort to combat cybercrime in the digital age.

Tactics in Phishing

Phishing attacks are a persistent and evolving threat that businesses face in today's digital landscape, leveraging a variety of deceptive tactics to exploit human vulnerabilities and gain unauthorized access to sensitive information. Understanding these common tactics is crucial for organizations to bolster their defenses and protect against the pervasive threat of phishing.[303] Here are some of the most prevalent tactics employed by cybercriminals in phishing attacks:

- **Email Spoofing**: One of the most common tactics used in phishing attacks is email spoofing, where attackers manipulate email header information to make messages appear as though they originate from a legitimate source.[304] By impersonating trusted entities, such as reputable companies or government agencies, attackers seek to deceive recipients into believing that the email is authentic and trustworthy.
- **Fake Websites:** Phishing attacks often involve the creation of fake websites that mimic legitimate web pages of well-known organizations or online services.[305] These spoofed websites are designed to trick users into entering their login credentials, financial information, or other sensitive data, which is then harvested by the attackers for malicious purposes.[306]
- **Social Engineering:** Social engineering techniques play a significant role in phishing attacks, as attackers exploit psychological principles and human behavior to manipulate victims into divulging confidential information or performing desired actions.[307] Tactics such as urgency, fear,

---

[303] Staff (2023). "The ripple effect: How one phishing attack can cause disaster across your organization." Cybsafe at https://www.cybsafe.com/blog/how-can-phishing-affect-a-business/.
[304] *Id.*
[305] United States Comptrollers of Currency Office (2024) at https://www.occ.gov/topics/consumers-and-communities/consumer-protection/fraud-resources/phishing-attack-prevention.html.
[306] *Id.*

curiosity, and authority are commonly employed to elicit emotional responses and prompt victims to act impulsively without questioning the legitimacy of the request.

- **Spear Phishing:** Spear phishing attacks target specific individuals or organizations with tailored and personalized messages that are carefully crafted to appear genuine and relevant to the recipient.[308] By leveraging information obtained from social media, company websites, or other sources, attackers can customize phishing emails to increase their likelihood of success and bypass traditional security measures.
- **Malicious Attachments:** Phishing emails often contain malicious attachments, such as infected documents or executable files, that are designed to exploit software vulnerabilities and compromise the victim's device. These attachments may contain malware, ransomware, or other malicious payloads that can be used to steal data, disrupt operations, or gain unauthorized access to systems.[309]
- **Link Manipulation:** Attackers frequently use link manipulation techniques to deceive users into clicking on malicious links that redirect them to fake websites or download malware onto their devices.[310] By disguising malicious URLs as legitimate ones or embedding them within seemingly innocuous content, attackers can trick users into unwittingly compromising their security.

---

[307] Kapersky Security at https://go.kaspersky.com/rs/802-IJN-240/images/Dangers_Phishing_Avoid_Lure_Cybercrime_ebook.pdf.
[308] University of Berkley (2024) Security office at https://security.berkeley.edu/faq/phishing/why-understanding-risk-phishing-important.
[309] OCC, *supra*
[310] *Id.*

- **Impersonation of Trusted Entities:** Phishing attacks often involve the impersonation of trusted entities, such as colleagues, supervisors, or IT administrators, to deceive recipients into complying with fraudulent requests.[311] By exploiting trust relationships and authority dynamics within organizations, attackers can trick employees into disclosing sensitive information or performing unauthorized actions.
- **Pretexting:** Pretexting is a form of social engineering where attackers fabricate elaborate scenarios or false pretenses to manipulate victims into divulging confidential information or providing access to restricted resources. By posing as legitimate individuals or authorities, attackers create a sense of urgency or importance that compels victims to comply with their demands.[312] This is covered in more detail in Chapter 13.

In essence, businesses must remain vigilant and proactive in defending against the diverse array of tactics employed in phishing attacks. By implementing robust security measures, conducting regular employee training and awareness programs, and leveraging advanced threat detection technologies, organizations can mitigate the risk of falling victim to phishing scams and protect their valuable assets, reputation, and customer trust from exploitation by cybercriminals.

Avoiding Phishing

To evade phishing attacks, companies must implement a multi-layered approach that combines technical safeguards,

---

311 Kapersky, *supra*.
312 *Id.*

user education, and proactive threat detection measures. Here are several methods that organizations can employ to mitigate the risk of falling victim to phishing attacks:

1. Email Filtering and Authentication: Implement robust email filtering solutions that can detect and block phishing emails before they reach users' inboxes.[313] Utilize techniques such as sender authentication protocols like SPF, DKIM, and DMARC to verify the legitimacy of incoming emails and identify potential spoofing attempts.
2. Employee Training and Awareness: Educate employees about the dangers of phishing attacks and provide regular training on how to recognize and report suspicious emails.[314] Teach employees to scrutinize email sender addresses, check for spelling and grammatical errors, and verify the legitimacy of requests for sensitive information or actions.
3. Multi-Factor Authentication (MFA): Require the use of multi-factor authentication for accessing sensitive systems and resources, especially for remote access or critical operations. MFA adds an extra layer of security by requiring users to provide additional verification, such as a one-time passcode or biometric authentication, in addition to their username and password.[315]
4. Secure Web Browsing: Encourage employees to use secure web browsers with built-in phishing protection features, such as real-time URL scanning and warning prompts for suspicious websites. Implement web filtering and URL categorization solutions to block access to known phishing sites and malicious content.[316]

---

[313] OCC, *Supra.*
[314] *Id.*
[315] *Id.*

5. Endpoint Security: Deploy advanced endpoint security solutions that include anti-malware, anti-phishing, and behavioral analysis capabilities to detect and prevent phishing attacks at the device level.[317] Implement regular software updates and patches to address security vulnerabilities and protect against exploit-based attacks.
6. Incident Response Planning: Develop comprehensive incident response plans that outline procedures for detecting, responding to, and mitigating phishing attacks in a timely manner.[318] Establish clear escalation paths, communication channels, and coordination mechanisms to facilitate rapid response and minimize the impact of security incidents.
7. Employee Reporting Mechanisms: Encourage employees to report suspected phishing emails or security incidents to the IT security team through designated reporting channels.[319] Implement an incident response hotline or email address where employees can quickly and confidentially report phishing attempts or suspicious behavior.
8. Ongoing Monitoring and Analysis: Continuously monitor network traffic, email logs, and system activity for signs of phishing activity or unauthorized access attempts.[320] Leverage security information and event management (SIEM) solutions and threat intelligence feeds to detect and analyze phishing-related indicators and proactively identify emerging threats.

---

[316] Kapersky, *supra.*
[317] CybSafe, *supra.*
[318] OCC, *supra.*
[319] *Id.*
[320] UC Berkley, *supra.*

By implementing these proactive measures and fostering a culture of security awareness and vigilance, companies can significantly reduce their susceptibility to phishing attacks and better protect their sensitive information, assets, and reputation from the pervasive threat posed by cybercriminals.

Conclusion

In conclusion, phishing attacks continue to pose significant threats to organizations worldwide, exploiting human vulnerabilities to gain unauthorized access to sensitive information and compromise corporate networks. Through a variety of deceptive tactics, cybercriminals exploit trust and manipulate unsuspecting users into divulging confidential credentials, clicking on malicious links, or downloading malware-infected attachments. The evolution of phishing techniques, including spear phishing, whaling, and pretexting, underscores the need for proactive measures to combat these sophisticated threats effectively.

To mitigate the risk of falling victim to phishing attacks, organizations must adopt a multi-faceted approach that combines technological defenses, user awareness training, and robust incident response capabilities. Implementing email filtering solutions, multi-factor authentication, and endpoint security measures can help detect and prevent phishing attempts before they cause harm. Equally important is fostering a security-conscious culture among employees, empowering them to recognize and report suspicious emails promptly.

Furthermore, organizations should regularly review and update their security policies and procedures to address emerging threats and evolving attack techniques. By staying informed about the latest phishing trends and sharing threat intelligence with industry peers, businesses can enhance their resilience against phishing attacks and better protect their valuable assets and reputation. Ultimately, combating phishing requires a collective effort involving proactive security measures, ongoing education, and continuous monitoring to stay one step ahead of cyber adversaries and safeguard critical business interests.

Chapter 12: Baiting Attacks

"And I think it's because good cons are all based on the victim's need, and the successful con artist is the one, I Guess, who can exploit that. I remember reading something about this, that one of the greatest tricks of confidence tricksters is the level that they flatter the victim."

Alfred Molina[321]

Chapter 12 looks into the insidious world of baiting attacks, a form of social engineering tactic where cybercriminals lure unsuspecting victims into compromising situations by offering enticing bait. Baiting attacks exploit human curiosity and greed, enticing individuals to click on malicious links, download infected files, or disclose sensitive information in exchange for promised rewards. This chapter explores the anatomy of baiting attacks, common methods employed by attackers, and strategies to mitigate the risks associated with such deceptive schemes.

Baiting attacks typically involve enticing victims with promises of desirable rewards, such as free software downloads, exclusive offers, or enticing content, to elicit a specific response. Attackers leverage various mediums, including email, social media, and physical devices like USB drives, to distribute bait and lure victims into taking the desired action.[322] Once the bait is taken, the attacker gains unauthorized access to the victim's system, steals sensitive information, or installs malware to compromise the target's network.

---

[321] (2024). "Con Man Quotes, Alfred Molina." AZ Quotes at https://www.azquotes.com/quotes/topics/con-artist.html.

[322] Staff (2022). "Web3 Security: Attach Types and Lessons Learned." Andreessen Howrowitz at https://a16z.com/web3-security-attack-types-and-lessons-learned/.

Common methods used in baiting attacks include the distribution of infected files disguised as legitimate software, enticing victims to click on malicious links embedded in enticing emails or social media posts, and leaving infected USB drives in public places for unsuspecting individuals to pick up and plug into their systems. Attackers may also impersonate trusted entities or individuals, such as IT support personnel or reputable organizations, to enhance the credibility of the bait and increase the likelihood of victim engagement.

To avoid falling prey to baiting attacks, individuals and organizations must exercise caution and adopt proactive security measures.[323] Implementing robust security policies and employee training programs to raise awareness about the dangers of baiting attacks and teach employees how to recognize and respond to suspicious emails, links, and files is crucial. Additionally, deploying advanced email filtering solutions, endpoint protection tools, and network security controls can help detect and block malicious content before it reaches end-users, reducing the risk of successful baiting attacks.

Organizations should enforce strict access controls and limit user privileges to prevent unauthorized installations of software or execution of potentially harmful files. Regularly updating software and operating systems with the latest security patches and fixes can also help mitigate vulnerabilities that attackers may exploit in baiting attacks. By fostering a security-conscious culture and encouraging employees to report suspicious activities promptly, businesses can strengthen their defenses against baiting attacks and protect sensitive information from falling into the wrong hands.

---

[323] *Id.*

Common Tactics within Methods

Baiting attacks come in various forms, each designed to exploit human psychology and lure unsuspecting victims into compromising situations. One common type of baiting attack is the distribution of infected files disguised as legitimate software or enticing content. Attackers may create malicious software packages masquerading as popular applications, games, or media files and distribute them through file-sharing platforms, email attachments, or compromised websites.[324] Unsuspecting users who download and execute these files unwittingly install malware onto their systems, allowing attackers to gain unauthorized access, steal sensitive information, or disrupt operations.

Another prevalent baiting tactic is the use of phishing emails or messages containing links to malicious websites. Attackers craft convincing emails that appear to originate from trusted entities, such as banks, government agencies, or reputable organizations, and lure recipients into clicking on embedded links or downloading attachments.[325] These links often lead to fake login pages designed to steal credentials, exploit vulnerabilities in web browsers or plugins, or initiate drive-by download attacks to deliver malware onto the victim's device. Similarly, phishing messages sent via SMS or social media platforms may contain enticing offers or urgent requests for action, aiming

---

[324] Staff (2022). "5 Types of Social Engineering attacks in Crypto and How to Prevent it." Lossless at https://losslessdefi.medium.com/5-types-of-social-engineering-attacks-in-crypto-and-how-to-prevent-it-4a669a3fef15.

[325] Poremba, Sue (2022). "The Dangerous Flaws of Web3 Security, According to a former Hacker." Security Intelligence at https://securityintelligence.com/articles/dangerous-flaws-web3-according-to-hacker/.

to deceive recipients into divulging personal information or performing specific actions that benefit the attacker.[326]

Physical baiting attacks involve the strategic placement of infected USB drives or other removable storage devices in public areas, such as parking lots, cafeterias, or conference rooms. These devices are often labeled with enticing labels, such as "Employee Bonuses" or "Confidential Data," to pique curiosity and encourage individuals to plug them into their computers.[327] Once connected, the malicious payloads hidden on these drives may automatically execute, infecting the victim's system with malware, or prompt the user to open files containing malware or phishing links. This method exploits human curiosity and the natural inclination to explore unknown content, making it an effective tactic for compromising systems and networks.

Another variant of baiting attacks involves the use of fake Wi-Fi hotspots or compromised public networks to intercept sensitive information transmitted over insecure connections.[328] Attackers may set up rogue access points in public spaces, such as airports, cafes, or hotels, with names similar to legitimate networks to lure unsuspecting users into connecting. Once connected, attackers can intercept and eavesdrop on network traffic, capturing login credentials, financial data, or other sensitive information transmitted by victims. Additionally, attackers may inject malicious content into web pages or redirect users to fake login portals to steal credentials or deliver malware to their devices.[329]

---

[326] *Id.*
[327] Andreessen Horowitz, *supra*.
[328] Keskin, Serkan (2023). "Digital Fortress, Part 2: How to security interact with Web3 Websites?" Coin Telegraph at https://cointelegraph.com/news/digital-fortress-part-2-how-to-securely-interact-with-web3-websites.

To avoid falling victim to baiting attacks, individuals and organizations must remain vigilant and adopt proactive security measures.[330] This includes educating employees and end-users about the risks associated with downloading or opening unfamiliar files, clicking on suspicious links, or connecting to unsecured networks. Implementing robust security controls, such as endpoint protection solutions, email filtering mechanisms, and network intrusion detection systems, can help detect and mitigate potential threats posed by baiting attacks. Additionally, organizations should enforce strict security policies regarding the use of external storage devices, access to sensitive networks, and the handling of confidential information to minimize the risk of compromise.[331] Regular security awareness training and simulated phishing exercises can further reinforce security awareness among employees, empowering them to recognize and report suspicious activities effectively.

How to Avoid Baiting

To mitigate the risk of falling victim to baiting attacks, companies must implement a multifaceted approach that encompasses technical controls, security awareness training, and robust policies and procedures. Here are several strategies companies can employ to avoid baiting attacks:

1. Employee Training and Awareness: Providing comprehensive security awareness training to employees is crucial for helping them recognize and respond appropriately to baiting attempts. Employees should be educated about the various

---

329 *Id.*
330 Staff (2024). "How to Avoid a Baiting Attack." RiskXChange at https://riskxchange.co/1006354/how-to-avoid-a-baiting-attack/.
331 *Id.*

types of baiting attacks, including phishing emails, malicious downloads, and fake Wi-Fi networks. Training sessions should emphasize the importance of exercising caution when encountering unfamiliar files, links, or devices and encourage employees to report suspicious activities promptly.

2. Strict Access Controls: Implementing stringent access controls can help limit the exposure of sensitive information and reduce the likelihood of unauthorized access. Companies should enforce the principle of least privilege, granting employees access only to the resources and systems necessary to perform their job responsibilities. Additionally, implementing strong authentication mechanisms, such as multi-factor authentication (MFA), can add an extra layer of security and prevent unauthorized access to corporate networks and systems.
3. Endpoint Protection: Deploying endpoint protection solutions, such as antivirus software, anti-malware scanners, and endpoint detection and response (EDR) tools, can help detect and prevent malicious activity on employee devices. These solutions can detect and block malicious files, scripts, and other payloads associated with baiting attacks before they can compromise the system. Regularly updating and patching endpoint security software ensures that devices are protected against the latest threats.
4. Email Filtering and Web Security: Implementing email filtering solutions and web security gateways can help block phishing emails and malicious websites associated with baiting attacks. These solutions use advanced threat intelligence and machine learning algorithms to analyze incoming emails and web traffic for signs of malicious activity, such as suspicious attachments, links, or content. By filtering out malicious content at the

email and web gateway, organizations can reduce the risk of employees falling victim to baiting attempts.

5. Secure File Sharing Practices: Establishing secure file sharing practices can help prevent employees from inadvertently downloading or sharing malicious files. Companies should encourage the use of centralized file storage and collaboration platforms equipped with built-in security features, such as encryption, access controls, and audit trails. Additionally, employees should be trained to verify the authenticity of file attachments and download links before opening or downloading them.
6. Network Segmentation and Monitoring: Segmenting the corporate network into distinct security zones and implementing network monitoring and intrusion detection systems can help detect and contain baiting attacks. By isolating critical systems and restricting lateral movement within the network, organizations can limit the impact of a successful baiting attack and prevent attackers from compromising sensitive data or resources. Continuous monitoring of network traffic and user behavior enables early detection of suspicious activity and timely response to potential security incidents.
7. Incident Response and Remediation: Establishing a comprehensive incident response plan that outlines procedures for detecting, responding to, and recovering from baiting attacks is essential for minimizing the impact of security incidents. Companies should conduct regular tabletop exercises and simulations to test the effectiveness of their incident response procedures and ensure that employees are prepared to respond effectively in the event of a security breach. Promptly investigating

and remediating security incidents can help prevent further damage and strengthen the organization's overall security posture.

8. Vendor Risk Management: Assessing and managing the security risks associated with third-party vendors and suppliers is critical for preventing baiting attacks originating from external sources. Companies should conduct thorough due diligence when selecting vendors and establish contractual agreements that include security requirements and obligations. Regularly auditing vendor security controls and monitoring third-party access to corporate systems and data can help mitigate the risk of supply chain attacks and ensure that vendors adhere to established security standards.[332]

Conclusion

Baiting attacks pose significant threats to organizations by exploiting human vulnerabilities and manipulating users into unwittingly compromising their security. These attacks rely on the promise of enticing rewards or the appearance of legitimacy to deceive victims and extract sensitive information or install malicious software. To defend against baiting attacks effectively, companies must adopt a proactive and multi-layered security approach that combines technical controls, employee awareness training, and robust policies and procedures.

By educating employees about the tactics used in baiting attacks and emphasizing the importance of vigilance and skepticism, organizations can empower their workforce to recognize and resist malicious baiting attempts. Implementing strict access controls, endpoint protection

---

[332] CISA, *supra.*

solutions, and email filtering mechanisms can help detect and block baiting attacks before they can inflict damage. Additionally, secure file sharing practices, network segmentation, and incident response planning are essential components of a comprehensive security strategy aimed at mitigating the risks posed by baiting attacks.

Fostering a culture of security awareness and accountability within the organization is crucial for promoting a collective defense against baiting attacks. Regular security training sessions, simulated phishing exercises, and incident response drills can help reinforce good security practices and prepare employees to respond effectively to potential threats. By taking proactive measures to prevent and mitigate baiting attacks, organizations can safeguard their sensitive data, protect their reputation, and maintain the trust of their customers and stakeholders in an increasingly hostile threat landscape.

Chapter 13: Pretexting

"When you realize how much of what you've constructed of yourself is based on deception and lies, that is a horrifying realization."

Jordan Peterson[333]

This chapter examines the deceptive tactic of pretexting, a form of social engineering where attackers fabricate a false scenario or identity to manipulate individuals into divulging sensitive information or performing actions against their interests. This chapter will explore the intricacies of pretexting, including its definition, prevalent tactics employed by attackers, and effective strategies to mitigate the risks associated with this deceptive practice.

Pretexting involves creating a pretext or plausible pretext to establish trust and credibility with the target, thereby lowering their guard and increasing the likelihood of successful manipulation.[334] Attackers often masquerade as trusted entities, such as colleagues, technical support personnel, or reputable organizations, to elicit sensitive information or gain unauthorized access to systems and data. By exploiting human emotions, social norms, and psychological vulnerabilities, pretexting schemes aim to deceive individuals into disclosing confidential information or performing actions that compromise security.

Common tactics used in pretexting include the use of pretext scripts or scenarios tailored to specific targets, manipulation of emotions such as fear or urgency to prompt immediate responses, and impersonation of authority

---

333 (2024). "Deception Quotes, Jordan Peterson." Brainy Quote at https://www.brainyquote.com/topics/deception-quotes.
334 Lossless, *Supra.*

figures or trusted contacts to lend credibility to the pretext.[335] Attackers may employ various communication channels, including phone calls, emails, social media messages, or in-person interactions, to execute their pretexting schemes and exploit unsuspecting victims.

To effectively mitigate the risks posed by pretexting attacks, organizations must implement robust security measures and cultivate a culture of awareness and skepticism among employees. This entails providing comprehensive training and education on recognizing and responding to pretexting attempts, including the importance of verifying the identity of individuals requesting sensitive information or access. Additionally, organizations should establish clear policies and procedures for handling requests for information or access, requiring verification of identities and authorization before disclosing sensitive data or performing actions.

Furthermore, leveraging technology solutions such as email filtering, caller ID verification, and multi-factor authentication can help detect and prevent pretexting attacks by enhancing the security posture of the organization's communication channels. Regular security assessments, simulated pretexting scenarios, and incident response drills are essential for evaluating the effectiveness of security controls and ensuring preparedness to respond to potential threats effectively.

By adopting a proactive approach to address pretexting risks, organizations can enhance their resilience against social engineering attacks and safeguard sensitive

---

[335] Sinegubko, Denis (2024). "Web3 Crypto Malware: Angel Drainer Overview, Variants & Stats." SucuriBlog at https://blog.sucuri.net/2024/02/web3-crypto-malware-angel-drainer.html.

information and assets from exploitation. By empowering employees with the knowledge and tools to identify and thwart pretexting attempts, organizations can effectively mitigate the risks posed by this deceptive tactic and protect their reputation, financial assets, and intellectual property from unauthorized access or disclosure.

Common Tactics in Pretexting

Pretexting, a deceptive tactic employed by malicious actors, encompasses various methods designed to manipulate individuals into divulging sensitive information or performing actions against their interests.[336] Understanding these common methods is crucial for companies to effectively identify and mitigate the risks associated with pretexting attacks.

One prevalent method of pretexting involves the creation of elaborate scenarios or stories to establish a false sense of urgency or importance.[337] Attackers may fabricate compelling narratives, such as urgent IT support requests, financial emergencies, or regulatory compliance audits, to coerce individuals into providing access credentials or confidential data.[338] By exploiting emotions like fear or concern, attackers aim to elicit immediate responses from their targets, bypassing their critical thinking and skepticism.

The next common tactic in pretexting is the impersonation of trusted entities or authority figures to gain the

---

[336] Staff (2024). "Web3 Trick or Treat? A practical guide to prevent phishing scams and hacks." Block Apex at https://medium.com/blockapex/web-3-trick-or-treat-a-practical-guide-to-prevent-phishing-scams-and-hacks-8e7d1fc29677.
[337] Lossless, *Supra*.
[338] *Id.*

confidence and cooperation of targets.[339] Attackers may pose as company executives, IT administrators, or law enforcement officials to lend credibility to their pretext and manipulate individuals into complying with their requests. By leveraging social engineering techniques and psychological manipulation, attackers exploit the inherent trust individuals place in authoritative or familiar figures, making them more susceptible to deception.

Pretexting attacks frequently involve the use of pretext scripts or prepared scenarios tailored to specific targets or situations.[340] These scripts provide attackers with a structured framework for engaging with their targets, guiding the conversation, and steering it toward the desired outcome. By rehearsing and refining their pretext narratives, attackers can effectively manipulate the conversation, extract sensitive information, or persuade targets to take actions that serve the attacker's objectives.[341]

Social media pretexting is another prevalent method wherein attackers gather personal information from publicly available sources, such as social media profiles, online forums, or company websites, to customize their pretext and establish rapport with targets.[342] By leveraging details about an individual's interests, affiliations, or relationships, attackers can craft convincing pretexts that resonate with their targets, increasing the likelihood of success. Social media pretexting underscores the importance of exercising caution when sharing personal information online and implementing privacy settings to limit exposure to potential attackers.

---

339 BlockApex, *supra.*
340 *Id.*
341 Lossless, *Supra.*
342 Sinegubko, *supra*

Pretexting attacks may exploit interpersonal relationships or organizational hierarchies to gain access to sensitive information or resources. Attackers may impersonate colleagues, vendors, or business partners to exploit trust relationships within an organization or exploit power differentials to coerce subordinates into complying with their demands.[343] By leveraging familiarity or authority dynamics, attackers can bypass traditional security measures and manipulate individuals into disclosing confidential information or granting unauthorized access.

Understanding the common methods of pretexting is essential for companies to effectively detect and mitigate the risks posed by social engineering attacks. By raising awareness among employees, implementing robust security protocols, and fostering a culture of skepticism and vigilance, organizations can fortify their defenses against pretexting and safeguard sensitive information from exploitation.

How to Avoid Pretexting

Pretexting, a deceptive tactic employed by malicious actors, encompasses various methods designed to manipulate individuals into divulging sensitive information or performing actions against their interests. Understanding these common methods is crucial for companies to effectively identify and mitigate the risks associated with pretexting attacks.

One prevalent method of pretexting involves the creation of elaborate scenarios or stories to establish a false sense of urgency or importance.[344] Attackers may fabricate

---

[343] *Id.*
[344] BlockApex, *supra.*

compelling narratives, such as urgent IT support requests, financial emergencies, or regulatory compliance audits, to coerce individuals into providing access credentials or confidential data. By exploiting emotions like fear or concern, attackers aim to elicit immediate responses from their targets, bypassing their critical thinking and skepticism.

Another common tactic in pretexting is the impersonation of trusted entities or authority figures to gain the confidence and cooperation of targets. Attackers may pose as company executives, IT administrators, or law enforcement officials to lend credibility to their pretext and manipulate individuals into complying with their requests.[345] By leveraging social engineering techniques and psychological manipulation, attackers exploit the inherent trust individuals place in authoritative or familiar figures, making them more susceptible to deception.

Additionally, pretexting attacks frequently involve the use of pretext scripts or prepared scenarios tailored to specific targets or situations. These scripts provide attackers with a structured framework for engaging with their targets, guiding the conversation, and steering it toward the desired outcome.[346] By rehearsing and refining their pretext narratives, attackers can effectively manipulate the conversation, extract sensitive information, or persuade targets to take actions that serve the attacker's objectives.

Social media pretexting is another prevalent method wherein attackers gather personal information from publicly available sources, such as social media profiles, online forums, or company websites, to customize their

---

[345] *Id.*
[346] Lossless, *Supra.*

pretext and establish rapport with targets.[347] By leveraging details about an individual's interests, affiliations, or relationships, attackers can craft convincing pretexts that resonate with their targets, increasing the likelihood of success. Social media pretexting underscores the importance of exercising caution when sharing personal information online and implementing privacy settings to limit exposure to potential attackers.

Furthermore, pretexting attacks may exploit interpersonal relationships or organizational hierarchies to gain access to sensitive information or resources.[348] Attackers may impersonate colleagues, vendors, or business partners to exploit trust relationships within an organization or exploit power differentials to coerce subordinates into complying with their demands.[349] By leveraging familiarity or authority dynamics, attackers can bypass traditional security measures and manipulate individuals into disclosing confidential information or granting unauthorized access.

In summary, understanding the common methods of pretexting is essential for companies to effectively detect and mitigate the risks posed by social engineering attacks. By raising awareness among employees, implementing robust security protocols, and fostering a culture of skepticism and vigilance, organizations can fortify their defenses against pretexting and safeguard sensitive information from exploitation.

---

347 *Id.*
348 CISA, *supra.*
349 *Id.*

Conclusion

Chapter 13 has broken down the concept of pretexting, shedding light on the deceptive tactics used by threat actors to manipulate individuals into divulging sensitive information or compromising security defenses. By exploring common methods of pretexting and strategies to mitigate associated risks, organizations can bolster their resilience against social engineering attacks and fortify their defenses.

Pretexting poses a significant threat to organizational security, exploiting human vulnerabilities to gain unauthorized access to sensitive information or resources. However, by understanding the tactics employed by pretexters and implementing proactive measures to counteract them, companies can effectively mitigate the risks posed by social engineering attacks.

Key takeaways from this chapter include the importance of employee training and awareness initiatives to cultivate a security-conscious culture, the implementation of robust verification and authentication protocols to validate identities and requests, the enforcement of confidentiality and data privacy policies to safeguard sensitive information, and the establishment of incident response procedures to detect and respond to pretexting attacks promptly.

In summary, addressing the challenges posed by pretexting requires a multifaceted approach that encompasses education, technology, and policy frameworks. By empowering employees with the knowledge and tools to recognize and thwart social engineering tactics, organizations can enhance their resilience to pretexting attacks and safeguard their assets and reputation in an increasingly complex threat landscape.

Chapter 14: Access Controls for Sensitive Information

"Knowledge is power. Information is power. The secreting or hoarding of knowledge or information may be an act of tyranny camouflaged as humility."

Robin Morgan[350]

Chapter 14 delves into access controls for sensitive information, highlighting the critical importance of security clearance, need-to-know principles, restricted access, and knowledge management strategies. These topics underscore the foundational elements of effective information security practices within organizations, ensuring that only authorized personnel can access and handle sensitive data.

Security clearance serves as a fundamental mechanism for regulating access to classified or sensitive information, establishing clear criteria and vetting processes to determine individuals' eligibility for accessing specific data sets or resources.[351] By rigorously evaluating employees' backgrounds, credentials, and trustworthiness, organizations can mitigate the risk of unauthorized disclosures or breaches of confidentiality.

Need-to-know principles dictate that access to sensitive information should be restricted to individuals with a legitimate business need for that information, minimizing exposure and limiting the potential impact of security incidents. This approach ensures that employees only

---

[350] (2024). "Access to Information. James Murdoch." AZ Quotes at https://www.azquotes.com/quotes/topics/access-to-information.html.
[351] Frank, Samuel (2023). "Web3 NFTs for Identity and Access Control, Now Available on Marketplace." Auth0 by Okta at https://auth0.com/blog/web3-nfts-for-identity-and-access-control/.

access data relevant to their roles or responsibilities, reducing the risk of inadvertent data exposure or misuse.

Restricted access and knowledge management mechanisms play a pivotal role in controlling access to sensitive information, employing encryption, authentication, and authorization protocols to enforce granular access controls and prevent unauthorized access or tampering. By implementing robust access control measures, organizations can safeguard sensitive data assets and prevent unauthorized disclosures or breaches.

Effective access control strategies involve a combination of technical controls, such as user authentication mechanisms, access control lists, and encryption, as well as administrative controls, such as policies, procedures, and training programs.[352] By adopting a layered approach to access control, organizations can create multiple barriers to unauthorized access and enhance their overall security posture.

Access controls for sensitive information are essential components of an organization's cybersecurity framework, ensuring that only authorized individuals can access and handle confidential or classified data. By implementing robust security clearance processes, adhering to need-to-know principles, and employing effective access control mechanisms, organizations can mitigate the risk of data breaches and protect their sensitive assets from unauthorized access or disclosure.

[352] TheNET, *supra*.

Security Clearance

Security clearance serves as a critical mechanism for determining who within an organization should be granted access to sensitive information.[353] The importance of security clearance lies in its ability to assess and verify individuals' trustworthiness, integrity, and suitability for accessing classified or confidential data. Here are five key reasons why security clearance is vital in deciding who should receive sensitive information:

1. Trustworthiness Assessment: Security clearance processes involve thorough background checks, including criminal history, financial records, and personal associations, to evaluate an individual's trustworthiness and reliability.[354] By scrutinizing an individual's past behavior and associations, organizations can assess the likelihood of unauthorized disclosures or breaches of confidentiality.
2. Risk Mitigation: Security clearance helps mitigate the risk of insider threats by identifying individuals who may pose a security risk due to their personal or professional backgrounds.[355] By screening employees and contractors for potential red flags or vulnerabilities, organizations can proactively address security risks and prevent insider attacks or data breaches.
3. Protection of Classified Information: Classified or sensitive information requires protection from

---

[353] Litan, Avivah (2023). "Web3 Blockchain enables users to take control of identity." Tech Target at https://www.techtarget.com/searchsecurity/post/Web3-blockchain-enables-users-to-take-control-of-identity.
[354] Frank, *supra.*
[355] TheNET, *supra.*

unauthorized access or disclosure to safeguard national security interests, intellectual property, or proprietary data.[356] Security clearance ensures that only individuals with a legitimate need for access to classified information are granted clearance, reducing the risk of espionage, sabotage, or data theft.

4. Access Control Enforcement: Security clearance establishes clear criteria and eligibility requirements for accessing classified or restricted data, enforcing access control policies and procedures within an organization.[357] By defining who can access specific types of information based on their clearance level, organizations can prevent unauthorized access and maintain the confidentiality and integrity of sensitive data assets.[358]
5. Compliance with Regulations: Many industries and government agencies have regulations and standards mandating security clearance for individuals handling sensitive or classified information.[359] By adhering to these regulatory requirements, organizations demonstrate their commitment to protecting sensitive data and ensuring compliance with legal and contractual obligations.

---

[356] Plumb, Taryn (2023). "Web3 Access Controls: How Zero Knowledge encryption can serve user access." Venture Beat at https://venturebeat.com/security/web3-access-controls-how-zero-knowledge-encryption-can-secure-user-access/.
[357] Thoams, Stanley (2023). "Authentication and Authorization in Web3: Securing the Decentralized Landscape." Security Token Offering at https://medium.com/security-token-offering/authentication-and-authorization-in-web3-securing-the-decentralized-landscape-10009b1e4215.
[358] Frank, *supra*.
[359] Litan, *supra*.

Security clearance plays a crucial role in determining who should receive access to sensitive information within an organization.[360] By assessing individuals' trustworthiness, mitigating security risks, protecting classified information, enforcing access control policies, and ensuring regulatory compliance, security clearance helps safeguard sensitive data assets and maintain confidentiality and integrity across the organization.

Neet to Know Information

Having specific systems in place for need-to-know information is crucial for maintaining confidentiality, limiting exposure to sensitive data, and mitigating the risk of unauthorized access or disclosure.[361] Here are four key reasons why implementing dedicated systems for need-to-know information is important:

1. Confidentiality Protection: Need-to-know information often pertains to classified, proprietary, or sensitive data that should only be accessed by individuals with a legitimate business need.[362] By segregating this information into dedicated systems with restricted access controls, organizations can prevent unauthorized personnel from inadvertently or intentionally accessing confidential data, reducing the risk of data leaks or breaches.
2. Access Control Enforcement: Dedicated systems for need-to-know information enable organizations to enforce granular access controls based on individuals' roles, responsibilities, and clearance levels.[363] By restricting access to authorized

[360] Plumb, *supra.*
[361] TheNET, *supra.*
[362] Frank, *supra.*
[363] Litan, *supra.*

personnel only, organizations can ensure that sensitive data remains protected from unauthorized viewing, modification, or dissemination, enhancing overall data security and integrity.

3. Minimization of Exposure: Need-to-know information is often distributed on a strictly need-to-know basis to minimize exposure and limit the potential impact of a security breach. By housing this information within dedicated systems with limited access, organizations can reduce the surface area for potential attacks and limit the scope of data exposure in the event of a security incident, thereby enhancing incident response capabilities and mitigating potential damage.
4. Compliance Requirements: Many industries and regulatory frameworks mandate the implementation of need-to-know principles and access controls to protect sensitive data and ensure compliance with data protection regulations.[364] By deploying dedicated systems for need-to-know information, organizations can demonstrate adherence to regulatory requirements, audit trails, and compliance standards, thereby avoiding potential penalties, fines, or legal liabilities associated with data breaches or non-compliance.

Having specific systems for need-to-know information is essential for safeguarding confidentiality, enforcing access controls, minimizing exposure, and meeting compliance requirements. By segregating sensitive data into dedicated systems with restricted access and implementing robust security measures, organizations can mitigate the risk of unauthorized access, protect sensitive information from

---

[364] TheNET, *supra.*

data breaches, and maintain compliance with regulatory mandates.

Restricted access and Knowledge

Restricting access to certain information, even for individuals with security clearance, is paramount for maintaining the principle of least privilege and ensuring that sensitive data remains protected from unauthorized disclosure or misuse. Here are three key reasons highlighting the importance of such access restrictions:

1. Principle of Least Privilege: The principle of least privilege dictates that individuals should only be granted access to the information and resources necessary to perform their job functions effectively.[365] By adhering to this principle, organizations can minimize the risk of data breaches or insider threats resulting from excessive permissions or privileges. Restricting access to certain information ensures that individuals with security clearance only have access to data that is directly relevant to their roles or responsibilities, reducing the likelihood of accidental or intentional misuse of sensitive information.
2. Defense-in-Depth Strategy: Implementing access restrictions as part of a defense-in-depth strategy adds an additional layer of security to protect sensitive data assets. Even individuals with security clearance may not require access to certain highly confidential or classified information unless it is essential for their specific tasks.[366] By segmenting data and enforcing access controls based on the

---

[365] Frank, *supra*.
[366] Litan, *supra*.

principle of least privilege, organizations can create multiple barriers to unauthorized access, making it more challenging for malicious actors to compromise sensitive data assets.

3. Mitigation of Insider Threats: Insider threats, whether unintentional or malicious, pose a significant risk to organizational security and data integrity. Restricting access to certain information helps mitigate the risk of insider threats by limiting the exposure of sensitive data to only those individuals who have a legitimate business need.[367] By carefully controlling access permissions and monitoring user activity, organizations can detect and respond to suspicious behavior or unauthorized access attempts, reducing the likelihood of data breaches or security incidents caused by insider threats.

Restricting access to certain information, even for individuals with security clearance, is essential for upholding the principle of least privilege, implementing a defense-in-depth strategy, and mitigating the risk of insider threats.[368] By carefully managing access permissions and enforcing access controls based on business need and role requirements, organizations can strengthen their overall security posture, protect sensitive data assets, and mitigate the potential impact of security breaches or insider incidents.

Conclusion

In conclusion, access controls play a crucial role in safeguarding sensitive information within organizations,

---

[367] *Id.*
[368] TheNET, *supra.*

ensuring that only authorized individuals have access to specific data based on their clearance levels, role requirements, and business needs. By implementing security clearance protocols, organizations can effectively determine who should be granted access to sensitive information, minimizing the risk of data breaches or unauthorized disclosures. Additionally, the concept of need-to-know information underscores the importance of restricting access to only those individuals who require specific data to perform their job functions, reducing the surface area for potential security threats.

The practice of restricting access to certain information, even for individuals with security clearance, serves as a fundamental principle of defense-in-depth security strategies. By creating multiple layers of security controls and enforcing access restrictions, organizations can strengthen their overall security posture and mitigate the risk of insider threats or unauthorized access attempts. Moreover, access controls help organizations adhere to regulatory compliance requirements and industry standards governing data protection and privacy, ensuring that sensitive information is handled in accordance with applicable laws and regulations.

In today's evolving threat landscape, where cyberattacks and insider threats pose significant risks to organizational security, effective access controls are essential for maintaining data confidentiality, integrity, and availability. By implementing robust access control mechanisms, organizations can reduce the likelihood of data breaches, protect sensitive information from unauthorized access or misuse, and safeguard their reputation and trustworthiness. Overall, access controls represent a critical component of comprehensive cybersecurity strategies, enabling organizations to proactively manage and mitigate security risks in an increasingly digital and interconnected environment.

Chapter 15: Endpoint Security Software

"Man has been reared by his errors: first he never saw himself other than imperfectly, second he attributed to himself imaginary qualities, third he felt himself in a false order of rank with animal and nature, fourth he continually invented new tables of values and for a time took each of them to be eternal and unconditional...If one deducts the effect of these four errors, one has also deducted away humanity, humaneness, and 'human dignity'"

Friedrich Nietzsche[369]

Chapter 15 breaks into endpoint security software, focusing on essential components such as strong firewalls, whitelists, exclusion protocols for unnecessary websites, and robust antivirus systems. These elements collectively form a comprehensive defense mechanism to safeguard endpoints against various cyber threats and malicious activities. By implementing robust endpoint security measures, organizations can effectively mitigate the risk of data breaches, malware infections, and unauthorized access attempts.

One of the key components of endpoint security is a strong firewall, which acts as a barrier between internal networks and external threats. Firewalls monitor and control incoming and outgoing network traffic, filtering out potentially harmful packets and preventing unauthorized access to sensitive data. By configuring firewalls to enforce strict access policies, organizations can fortify their network perimeter and prevent cyber attackers from exploiting vulnerabilities to gain access to critical systems and information.

[369] (2024). "Human Error, Friedrich Nietzsche." Good Reads at https://www.goodreads.com/quotes/tag/human-error.

Whitelists are another critical aspect of endpoint security, allowing organizations to specify which applications, websites, or services are permitted to access their network resources. By creating whitelists of trusted entities, organizations can reduce the attack surface and minimize the risk of unauthorized access or infiltration by malicious actors. Additionally, whitelisting helps organizations enforce security policies and compliance requirements, ensuring that only authorized software and services are allowed to interact with endpoint devices.

Exclusion protocols for unnecessary websites, such as social media platforms and non-work-related sites, are essential for maintaining productivity and minimizing distractions in the workplace. By implementing policies that restrict access to non-essential websites, organizations can prevent employees from inadvertently accessing malicious or inappropriate content that could compromise endpoint security. Moreover, exclusion protocols help organizations conserve bandwidth and network resources by limiting access to websites that are not essential for business operations.

Robust antivirus systems play a crucial role in detecting and mitigating malware threats on endpoint devices, including computers, laptops, and mobile devices. Antivirus software continuously monitors system activities, scans for known malware signatures, and blocks malicious files or applications from executing on endpoints. By deploying advanced antivirus solutions with real-time threat detection capabilities, organizations can proactively identify and neutralize malware infections before they cause significant damage or data loss.

Overall, endpoint security software serves as a critical line of defense against cyber threats targeting endpoint devices within organizations. By implementing strong firewalls, whitelists, exclusion protocols, and robust antivirus systems, organizations can strengthen their endpoint security posture, protect sensitive data, and mitigate the risk of cyber attacks and data breaches. Additionally, proactive endpoint security measures help organizations maintain regulatory compliance, safeguard their reputation, and preserve business continuity in the face of evolving cyber threats.

Strong Firewalls

Firewalls serve as a critical component of network security, acting as a barrier between internal networks and external threats.[370] Strong firewalls are essential for organizations to control and monitor incoming and outgoing network traffic effectively. They analyze data packets and enforce access policies to prevent unauthorized access and block malicious activities.

Firstly, strong firewalls protect organizations' network perimeters by filtering incoming traffic and blocking suspicious or malicious packets.[371] By inspecting packets at the network level, firewalls can identify and block potentially harmful threats before they reach internal systems, reducing the risk of cyber attacks and data breaches.

Secondly, firewalls help organizations enforce security policies and compliance requirements by regulating network access based on predefined rules and criteria.[372]

---

[370] Staff (2024). "What is a Firewall?" Checkpoint at https://www.checkpoint.com/cyber-hub/network-security/what-is-firewall/.
[371] *Id.*
[372] Staff (2024). "Your Web3 Firewall." Zengo at

They can restrict access to specific ports, protocols, or IP addresses, ensuring that only authorized traffic is allowed to enter or leave the network.[373]

Thirdly, strong firewalls provide organizations with granular control over network traffic, allowing them to create custom rules and configurations tailored to their specific security needs.[374] Administrators can define access policies based on factors such as user roles, device types, and geographic locations, enabling fine-grained control over network access and reducing the attack surface.

Lastly, strong firewalls play a crucial role in protecting against emerging threats and zero-day vulnerabilities by continuously updating their threat intelligence databases and signature libraries.[375] They can detect and block new and evolving threats in real-time, providing organizations with proactive defense capabilities against the latest cyber attacks and malware variants.

Whitelists

Whitelists are an essential component of endpoint security, allowing organizations to specify which applications, websites, or services are permitted to access their network resources.[376] Strong whitelists help organizations reduce the attack surface and minimize the risk of unauthorized access or infiltration by malicious actors.

---

https://zengo.com/firewall/.
[373] Staff (2023). "How do web3 firewalls and smart contract security services mitigate crypto security flaws." Gate Blog at https://www.gate.io/blog_detail/1738/how-do-web3-firewalls-and-smart-contract-security-services-mitigate-crypto-security-flaws.
[374] Zengo, *supra*.
[375] *Id.*
[376] Metaschool, *supra*.

Firstly, strong whitelists enable organizations to enforce strict access controls and prevent unauthorized applications or services from executing on endpoint devices.[377] By creating comprehensive whitelists of trusted entities, organizations can block potentially harmful or unauthorized software from running, mitigating the risk of malware infections and data breaches.

Secondly, whitelists help organizations maintain regulatory compliance and enforce security policies by ensuring that only approved software and services are allowed to interact with endpoint devices.[378] By defining clear guidelines for acceptable software usage, organizations can minimize the risk of non-compliance and ensure adherence to industry regulations and standards.[379]

Thirdly, strong whitelists provide organizations with visibility and control over endpoint activities, allowing them to monitor and manage application usage effectively.[380] Administrators can track the execution of whitelisted applications, identify unauthorized software installations, and take appropriate action to remediate security incidents or policy violations.

Lastly, strong whitelists contribute to improved endpoint performance and productivity by minimizing the impact of unauthorized or non-essential applications on system resources.[381] By restricting access to approved software and services, organizations can optimize endpoint performance, reduce system downtime, and enhance user productivity.

---

377 *Id.*
378 Web3auth at https://web3auth.io/docs/dashboard-setup/whitelisting.
379 Staff (2024). "Whitelists." Whitelist Zone at https://wzone.io/.
380 *Id.*
381 Web3auth, *supra.*

Strong firewalls and whitelists are essential components of a comprehensive cybersecurity strategy, helping organizations protect their networks and endpoints from a wide range of threats and vulnerabilities. By implementing robust firewalls and whitelists, organizations can strengthen their security posture, minimize the risk of cyber attacks and data breaches, and maintain regulatory compliance and business continuity.

Exclusion of Unnecessary Websites

Maintaining a list of excluded websites, particularly those associated with social media platforms and explicit content such as pornography, is crucial for enhancing cybersecurity within organizations. Firstly, blocking access to social media websites helps mitigate the risk of employees falling victim to social engineering attacks and phishing scams, which are often perpetrated through social media platforms.[382] By restricting access to these sites, organizations can minimize distractions and improve productivity while reducing the likelihood of employees inadvertently disclosing sensitive information or engaging in risky online behavior.

Secondly, excluding websites containing explicit or inappropriate content, such as pornography, is essential for creating a safe and professional work environment.[383] Accessing such content not only poses a security risk due to potential malware infections but also creates legal and ethical concerns for organizations. By implementing web filtering mechanisms to block access to inappropriate websites, organizations can uphold their corporate policies,

---

[382] GitHub at https://github.com/scamsniffer/scam-database.
[383] Weston, Georgia (2023). "List of top 5 Web3 Security Tools." 101 Blockchains at https://101blockchains.com/top-web3-security-tools/.

maintain employee morale, and mitigate the risk of legal liabilities arising from inappropriate online activities.

Thirdly, restricting access to certain websites helps conserve network bandwidth and optimize internet usage within organizations.[384] Social media platforms and streaming websites consume significant network resources, potentially impacting the performance of critical business applications and services. By excluding non-essential websites from the network, organizations can allocate bandwidth more effectively, improve network performance, and ensure a smoother user experience for legitimate business activities.

Antivirus Systems

Deploying a robust antivirus system is paramount for safeguarding organizational assets and data against a wide range of cyber threats, including viruses, malware, ransomware, and spyware. Firstly, antivirus software serves as a critical defense mechanism against malicious software by detecting and removing malicious code from endpoint devices. It continuously monitors system activities, files, and network traffic for signs of suspicious behavior, providing real-time protection against known and emerging threats.

Secondly, a strong antivirus system helps prevent unauthorized access and data breaches by detecting and blocking malicious attacks aimed at compromising sensitive information. By scanning files and email attachments for malware signatures and malicious behaviors, antivirus software can thwart cybercriminals' attempts to infiltrate organizational networks and steal

[384] Frontal Blacklist archives at https://frontal.io/tag/blacklist/.

confidential data. Additionally, proactive threat detection and response capabilities enable antivirus solutions to quarantine or neutralize threats before they can cause significant harm, minimizing the impact of cyber-attacks on business operations.

Thirdly, antivirus software plays a crucial role in maintaining regulatory compliance and data privacy standards by protecting sensitive information from unauthorized access or disclosure. Compliance mandates such as the General Data Protection Regulation (GDPR) and the Health Insurance Portability and Accountability Act (HIPAA) require organizations to implement appropriate security measures to safeguard personal and sensitive data.[385] By deploying a strong antivirus solution, organizations can demonstrate due diligence in protecting customer data, mitigating the risk of regulatory fines, legal penalties, and reputational damage associated with data breaches.

Having a comprehensive approach to cybersecurity that includes the exclusion of certain websites and the deployment of a strong antivirus system is essential for protecting organizational assets, maintaining productivity, and ensuring regulatory compliance. By implementing web filtering mechanisms to restrict access to non-essential websites and deploying robust antivirus software to detect and neutralize cyber threats, organizations can strengthen their security posture, mitigate the risk of cyber-attacks and data breaches, and safeguard sensitive information from unauthorized access or disclosure.

---

[385] Intersoft (2023). "General Data Protection Regulation." At https://gdpr-info.eu/.

Conclusion

In conclusion, robust endpoint security software comprising strong firewalls, whitelists, exclusion protocols, and antivirus systems is indispensable for safeguarding organizational assets and data against a myriad of cyber threats. By implementing these comprehensive security measures, organizations can fortify their defenses, mitigate the risk of cyber attacks, and enhance overall cybersecurity posture. Moreover, endpoint security software plays a pivotal role in maintaining regulatory compliance, protecting sensitive information, and upholding the trust of customers and stakeholders.

The proactive detection and response capabilities of endpoint security solutions enable organizations to swiftly identify and neutralize emerging threats, minimizing the potential impact of cyber-attacks on business operations. By continuously monitoring endpoint devices and network traffic for signs of malicious activity, organizations can thwart cybercriminals' attempts to infiltrate their networks, steal confidential data, or disrupt critical systems. Additionally, endpoint security software helps organizations adapt to evolving cyber threats and emerging attack vectors, ensuring resilience against sophisticated cyber attacks.

The integration of endpoint security software with centralized management platforms enables organizations to streamline security operations, automate threat detection and response processes, and enhance visibility into their digital environments. By consolidating security controls and management functions within a unified platform, organizations can improve operational efficiency, reduce complexity, and achieve greater control over their security posture. Overall, investing in robust endpoint security software is imperative for organizations seeking to mitigate cyber risks, protect critical assets, and uphold the integrity and confidentiality of their data in an increasingly interconnected and digitally-driven world.

Chapter 16: Third Party Contractor Management

"Safety and security don't just happen, they are the result of collective consenss and public investment. We owe our children, the most vulnerable citizens in our society, a life free of violence and fear."

Nelson Mandela[386]

In Chapter 16, we examine the critical aspects of managing third-party contractors to ensure robust cybersecurity measures and protect organizational assets. This entails conducting comprehensive security assessments for contractors, drafting robust contracts, implementing continuous monitoring mechanisms, and establishing indemnification agreements to mitigate potential risks and liabilities.

Security assessments for contractors serve as an essential first step in evaluating the cybersecurity posture and practices of third-party entities before engaging in business partnerships or collaborations. By conducting thorough assessments, organizations can identify any vulnerabilities, weaknesses, or deficiencies in the contractor's security infrastructure and practices, thereby enabling informed decision-making regarding risk management and mitigation strategies.

Equally important is the drafting of well-defined and legally binding contracts that outline the rights, responsibilities, obligations, and expectations of both parties concerning cybersecurity practices and data

[386] (2024). "Safety and Security Quotes, Nelson Mandela." AZ Quotes at https://www.azquotes.com/quotes/topics/safety-and-security.html.

protection measures. These contracts should include specific clauses pertaining to cybersecurity requirements, data handling protocols, incident response procedures, and compliance with regulatory standards to ensure alignment with organizational security policies and regulatory obligations.

Continuous monitoring of third-party contractors is essential to maintain visibility into their cybersecurity posture and detect any deviations or anomalies that may pose security risks to the organization's data and infrastructure. By implementing robust monitoring mechanisms and leveraging security analytics tools, organizations can proactively identify and address potential security threats or breaches arising from contractor activities or interactions.

Establishing indemnification agreements that outline liability provisions and mechanisms for addressing losses incurred due to contractor negligence or breaches of contractual obligations is crucial for protecting the organization's interests and mitigating financial risks. These agreements should clearly delineate the extent of liability, the process for seeking compensation, and any recourse available in the event of security incidents or data breaches attributable to contractor actions.

In essence, effective third-party contractor management requires a comprehensive approach encompassing security assessments, contractual agreements, continuous monitoring, and indemnification mechanisms to mitigate cybersecurity risks, safeguard organizational assets, and uphold the integrity and confidentiality of sensitive data. By implementing these measures, organizations can foster stronger partnerships with third-party contractors while ensuring robust cybersecurity posture and regulatory compliance.

Security Assessments

Security assessments of third-party contractors are crucial for safeguarding organizational assets and mitigating potential cybersecurity risks. These assessments provide insights into the contractor's security posture, including their infrastructure, policies, and practices, helping organizations identify vulnerabilities and weaknesses that may pose threats to sensitive data and critical systems.[387] By conducting thorough assessments, organizations can make informed decisions about engaging with contractors, ensuring they meet the necessary security standards and compliance requirements.

Security assessments enable organizations to assess the level of risk associated with third-party engagements and develop risk mitigation strategies accordingly. By identifying and prioritizing risks, organizations can allocate resources effectively to address critical security issues and implement appropriate controls to mitigate potential threats.[388] Additionally, security assessments help establish a baseline for ongoing monitoring and evaluation of contractor performance, allowing organizations to track improvements over time and ensure continued compliance with security requirements.

---

[387] Deloitte, *supra*.

[388] Weston, Georgia (2023B). "10 top web3 security startups and companies." 101 blockchains at https://101blockchains.com/top-web3-security-startups-and-companies/. *See Also*. Staff (2024). "Web3 Penetration Testing." Cert Cube Labs at https://certcube.com/web3-penetration-testing-solutions/.

Security assessments enhance transparency and accountability in third-party relationships by fostering open communication and collaboration between organizations and contractors.[389] By engaging in dialogue about security expectations and requirements upfront, organizations can establish mutual understanding and commitment to cybersecurity best practices, laying the foundation for a strong and secure partnership. Overall, security assessments play a vital role in promoting trust, confidence, and resilience in third-party relationships, ultimately strengthening the organization's overall cybersecurity posture.

Contractual Agreements

Ironclad contractor agreements are essential for defining the rights, responsibilities, and obligations of both parties involved in a business relationship. These agreements establish clear expectations regarding cybersecurity practices, data handling procedures, and compliance requirements, ensuring alignment with organizational policies and regulatory standards. By outlining specific contractual terms and conditions, organizations can mitigate legal and financial risks associated with third-party engagements.

Ironclad contractor agreements help mitigate potential disputes or misunderstandings by providing a legally binding framework for addressing issues related to security breaches, data breaches, or non-compliance with contractual obligations. By clearly defining the consequences of breach of contract and establishing mechanisms for dispute resolution, these agreements enable

---

[389] CISA, *supra.*

organizations to enforce accountability and protect their interests in the event of contractual disputes.

Additionally, ironclad contractor agreements serve as a valuable tool for managing expectations and setting performance standards for third-party contractors. By detailing the scope of work, deliverables, timelines, and service level agreements, these agreements help ensure that contractors understand their roles and responsibilities and are held accountable for meeting contractual obligations. Furthermore, by including provisions for periodic reviews and performance evaluations, organizations can monitor contractor performance and address any issues or concerns proactively.

Overall, ironclad contractor agreements provide a solid legal framework for governing third-party relationships, protecting organizational assets, and mitigating risks associated with outsourcing activities. By establishing clear guidelines and expectations upfront, organizations can foster trust, transparency, and compliance in their interactions with contractors, ultimately enhancing the security and integrity of their operations.

Continuous Monitoring

Continuous monitoring of contractors is essential for maintaining a proactive approach to cybersecurity risk management and ensuring ongoing compliance with security requirements. By continuously monitoring contractor activities, organizations can detect and respond to potential security incidents or anomalies in real-time, minimizing the impact of breaches and preventing unauthorized access to sensitive data.[390] Additionally,

[390] Staff (2023). "What is On-Chain Monitoring." AuditOne at

continuous monitoring allows organizations to track changes in the contractor's security posture over time, identify emerging threats or vulnerabilities, and implement timely remediation measures to mitigate risks effectively.

Furthermore, continuous monitoring enables organizations to maintain visibility and oversight of contractor activities throughout the duration of the engagement, ensuring adherence to contractual obligations and security policies. By monitoring access logs, network traffic, and system activities, organizations can identify unauthorized behavior or suspicious activities that may indicate potential security breaches or compliance violations.[391] Moreover, continuous monitoring facilitates regular performance evaluations and compliance audits, providing insights into the contractor's adherence to security standards and regulatory requirements and enabling organizations to address any deficiencies or shortcomings promptly.

Overall, continuous monitoring of contractors helps organizations proactively manage cybersecurity risks, strengthen their security posture, and protect against potential threats to sensitive information and critical systems. By maintaining vigilance and oversight over contractor activities, organizations can ensure the integrity, confidentiality, and availability of their data and systems, ultimately safeguarding their business operations and reputation from the adverse impacts of security breaches or compliance failures.

---

https://medium.com/@auditone.io/what-is-on-chain-monitoring-f792ffd6e8a0.
[391] CISA, *supra*.

Indemnification Agreements

Indemnification agreements are essential legal instruments that protect organizations from financial liabilities arising from third-party contractor negligence or breach of contract. These agreements stipulate that the contractor will indemnify the organization against any losses, damages, or expenses incurred as a result of the contractor's actions or failures to meet contractual obligations, including breaches of data security or privacy breaches. By including indemnification provisions in contractor agreements, organizations can transfer the financial burden of potential security incidents or data breaches to the contractor, mitigating their exposure to legal and financial risks.

Indemnification agreements provide organizations with recourse in the event of a security breach or data incident involving the contractor's actions or negligence. By securing indemnification from the contractor, organizations can recover damages or expenses associated with breach response, forensic investigations, regulatory fines, legal fees, and remediation efforts, minimizing the financial impact of security incidents on their operations and bottom line.[392] Additionally, indemnification agreements serve as a deterrent against contractor misconduct or negligence by holding them accountable for the consequences of their actions and incentivizing compliance with security standards and contractual obligations.

Furthermore, indemnification agreements enhance the organization's ability to enforce contractual requirements and protect its interests in third-party relationships. By establishing clear provisions for indemnification and liability allocation, these agreements ensure that both

---

[392] *Id.*

parties understand their respective responsibilities and obligations regarding cybersecurity and data protection. Moreover, indemnification agreements provide organizations with legal recourse in the event of disputes or breaches of contract, enabling them to seek compensation for damages or losses incurred due to contractor negligence or non-compliance with security requirements.

Overall, indemnification agreements play a crucial role in mitigating legal and financial risks associated with third-party engagements, providing organizations with financial protection and recourse in the event of security incidents or breaches involving contractors. By securing indemnification provisions and enforcing contractual obligations, organizations can safeguard their interests, protect against potential liabilities, and ensure accountability and compliance in their relationships with third-party contractors.

Conclusion

Chapter 16 considers the critical aspects of third-party contractor management in the realm of cybersecurity. It underscores the indispensable need for robust security assessments, ironclad contractual agreements, continuous monitoring, and comprehensive indemnification agreements when engaging with external contractors. These components collectively form a formidable defense mechanism against potential security risks and liabilities arising from third-party engagements.

By emphasizing the importance of stringent security assessments, organizations can effectively evaluate the cybersecurity posture of their contractors, identify potential vulnerabilities, and mitigate risks before they escalate into major security incidents. Additionally, the implementation of thorough contractual agreements ensures that both parties clearly delineate their roles, responsibilities, and obligations concerning cybersecurity, thereby establishing a framework for accountability and compliance throughout the engagement.

Continuous monitoring emerges as a crucial element in maintaining ongoing oversight and vigilance over contractor activities, enabling organizations to detect and respond promptly to any anomalies or security breaches. Through continuous monitoring, organizations can proactively address emerging threats, enforce compliance with security policies, and uphold the integrity and confidentiality of their data and systems. Furthermore, the inclusion of comprehensive indemnification agreements provides organizations with a layer of financial protection and recourse in the event of contractor negligence or breaches of contract, minimizing the potential impact of security incidents on their operations and reputation.

In conclusion, effective third-party contractor management is essential for safeguarding organizational assets, mitigating cybersecurity risks, and ensuring regulatory compliance in an increasingly interconnected business landscape. By adopting a proactive approach to contractor management, organizations can strengthen their cybersecurity posture, foster trust and confidence in their business relationships, and mitigate the adverse impacts of security breaches and compliance failures.

Conclusion Part III

Part III of this book has provided a comprehensive exploration of social engineering and phishing, emphasizing the importance of building a culture of security within organizations to effectively combat these pervasive threats. As businesses navigate an increasingly digital landscape fraught with sophisticated cyberattacks and data breaches, fostering a culture of security becomes paramount in mitigating risks, protecting sensitive information, and preserving organizational integrity.

At its core, a culture of security instills a mindset of vigilance, accountability, and responsibility among employees, empowering them to become active participants in safeguarding organizational assets and thwarting malicious activities. By prioritizing security awareness and education initiatives, organizations can equip their workforce with the knowledge, skills, and resources necessary to recognize and respond effectively to evolving cyber threats, thereby reducing the likelihood of falling victim to social engineering attacks and phishing scams.

Cultivating a culture of security involves promoting a sense of ownership and collective responsibility for cybersecurity across all levels of the organization, from frontline employees to executive leadership. By fostering open communication channels, encouraging reporting of suspicious activities, and establishing clear protocols for incident response, organizations can create an environment where security is woven into the fabric of everyday operations and decision-making processes.

Furthermore, a culture of security extends beyond individual behavior to encompass organizational policies, procedures, and technologies designed to fortify defenses and mitigate vulnerabilities. By implementing robust access controls, encryption measures, and multi-factor authentication systems, organizations can bolster their resilience against social engineering tactics and phishing attacks, reducing the likelihood of unauthorized access to sensitive information and data breaches.

Central to building a culture of security is the recognition that cybersecurity is not solely the responsibility of the IT department but a shared commitment that permeates every aspect of the organization. By fostering collaboration and cross-functional partnerships, organizations can leverage diverse expertise and perspectives to develop holistic security strategies that address emerging threats and align with business objectives.

Embracing a culture of security requires a commitment to ongoing evaluation, adaptation, and improvement in response to evolving cyber threats and organizational needs. By conducting regular security assessments, audits, and training exercises, organizations can identify areas for enhancement, address vulnerabilities, and refine their security posture to stay ahead of emerging threats.

In summary, building a culture of security is imperative for organizations seeking to safeguard their assets, maintain customer trust, and thrive in an increasingly digital and interconnected world. By fostering a shared commitment to security, investing in education and awareness, and implementing robust policies and technologies, organizations can empower their workforce to become proactive defenders against social engineering and phishing attacks, thereby fortifying their defenses and preserving their reputation and resilience in the face of cyber threats.

Part IV: Physical Security

Introduction Part IV

Part IV of our comprehensive exploration explores the issues of physical security within cybersecurity in regard to the context of web3 enterprises. As technology continues to evolve at a rapid pace, so too do the threats and challenges facing organizations in safeguarding their digital assets and operations. In this section, we dissect various facets of cybersecurity, ranging from hardware security to insurance coverage, each playing a crucial role in fortifying the defenses of web3 companies against an array of cyber threats.

Chapter 17 sets the stage by examining the critical importance of hardware security in the web3 era. With storage devices and wallets serving as gateways to sensitive digital assets, ensuring their secure storage and protection becomes paramount. We delve into the nuances of tamper-resistant hardware and the secure disposal of outdated devices, highlighting the need for robust measures to safeguard against physical and digital intrusions.

Moving forward, Chapter 18 shines a spotlight on data center security, recognizing data centers and server farms as the backbone of web3 operations. From controlling physical access to implementing environmental controls, this chapter emphasizes the need for comprehensive strategies to mitigate risks associated with both external threats and internal vulnerabilities. Furthermore, the importance of disaster recovery plans and proper document disposal practices underscores the proactive measures required to safeguard sensitive data in the event of unforeseen contingencies.

In Chapter 19, the focus shifts to employee security, acknowledging the pivotal role of human factors in cybersecurity. Through background checks, security awareness training, and policy implementation, organizations can empower employees to become the first line of defense against cyber threats. By fostering a culture of vigilance and accountability, web3 companies can enhance their resilience against social engineering tactics and insider threats.

Chapter 20 extends the scope of cybersecurity to encompass supply chain management, recognizing the interconnected nature of modern business ecosystems. From vendor risk assessments to secure delivery facilities, this chapter underscores the importance of vetting and monitoring third-party partners to mitigate supply chain risks and vulnerabilities. Moreover, robust contractual agreements and insurance coverage further fortify the defenses of web3 companies against potential disruptions and liabilities.

In Chapter 21, we delve into the realm of event security, acknowledging the unique challenges posed by corporate gatherings and industry conferences. With corporate espionage and social engineering tactics prevalent in such environments, implementing stringent security protocols becomes imperative. From employee training to attendee vetting, this chapter outlines proactive measures to safeguard sensitive information and mitigate risks associated with public engagements.

Finally, Chapter 22 addresses the role of insurance coverage in bolstering cybersecurity resilience within web3 enterprises. From corporate insurance policies to errors and omissions coverage, this chapter underscores the importance of risk mitigation strategies in safeguarding against financial losses and liabilities arising from cyber incidents. By embracing comprehensive insurance coverage, web3 companies can navigate the complex landscape of cybersecurity risks with confidence and resilience.

In essence, Part IV provides a comprehensive roadmap for web3 companies seeking to fortify their cybersecurity defenses and navigate the evolving threat landscape with confidence and resilience. Through proactive measures, robust protocols, and strategic partnerships, organizations can safeguard their digital assets and operations against an array of cyber threats, ensuring continuity and success in the dynamic digital ecosystem of the web3 era.

Chapter 17: Hardware Security

"Hardware is easy to protect: lock it in a room, chain it to a desk, or buy a spare. Information poses more of a problem. It can exist in more than one place; be transported halfway across the planet in seconds; and be stolen without your knowledge."

Bruce Schneier[393]

In the ever-evolving landscape of cybersecurity, one aspect that demands utmost attention is hardware security. As companies embrace the decentralized paradigm of web3 technologies, the integrity and protection of hardware devices become paramount. Chapter 17 delves into the intricate realm of hardware security, addressing key concerns and best practices for safeguarding sensitive data and assets.

At the core of web3 operations lie storage devices and wallets, serving as the custodians of digital assets and cryptographic keys. These devices play a pivotal role in securing private keys, facilitating transactions, and managing decentralized identities. However, they also present inherent vulnerabilities, making them susceptible to various attacks, including physical tampering and remote exploits. This section examines strategies for enhancing the security posture of storage devices and wallets, ranging from robust encryption mechanisms to multi-factor authentication protocols.

The advent of tamper-resistant hardware marks a significant milestone in fortifying the security of critical

---

[393] (2024). "Hardware Security, Bruce Schneier." Quote Fancy at https://quotefancy.com/quote/1570643/Bruce-Schneier-Hardware-is-easy-to-protect-lock-it-in-a-room-chain-it-to-a-desk-or-buy-a.

systems and infrastructure. Designed to withstand physical attacks and unauthorized access attempts, tamper-resistant hardware employs advanced techniques such as secure enclaves, hardware-based attestation, and secure boot mechanisms. This chapter explores the principles behind tamper-resistant design and evaluates its effectiveness in thwarting sophisticated adversaries' exploitation attempts.

As hardware components reach the end of their operational lifecycle, ensuring their secure disposal becomes imperative to prevent data breaches and unauthorized access. Improper disposal practices pose significant risks, as residual data remnants may be exploited by malicious actors to glean sensitive information. By implementing secure disposal procedures, organizations can mitigate these risks and uphold data privacy and integrity standards. This section offers insights into best practices for securely decommissioning outdated hardware and sanitizing storage media to render data irretrievable.

In an era defined by digital transformation and decentralized technologies, the integrity of hardware devices underpins the foundation of cybersecurity. Chapter 17 examines the intricacies of hardware security, illuminating the critical role of storage devices, tamper-resistant hardware, and secure disposal practices in safeguarding web3 ecosystems. By embracing proactive measures and robust security protocols, organizations can fortify their hardware infrastructure and mitigate emerging threats in the dynamic landscape of cybersecurity.

Secure Storage of Wallets and Devices

Securing storage devices and wallets is paramount for ensuring the robustness and resilience of web3 businesses. These devices serve as repositories for cryptographic keys, which are fundamental to accessing and managing digital assets on blockchain networks.[394] Any compromise to the security of these keys can result in unauthorized access to sensitive information, leading to financial losses, data breaches, and reputational damage. Therefore, by employing stringent security measures such as encryption, multi-factor authentication, and hardware-based security modules, businesses can fortify their storage infrastructure and safeguard against unauthorized access and malicious attacks.

Second, in the decentralized landscape of web3, where individuals have full control over their digital identities and assets, the integrity of storage devices directly impacts user trust and confidence. Users entrust web3 businesses with the custody of their private keys and personal data, expecting robust security measures to protect their assets from unauthorized access and exploitation. Secure storage solutions instill confidence among users, fostering long-term relationships and enhancing the reputation of web3 businesses as reliable custodians of digital assets.[395] By prioritizing secure storage practices, businesses can uphold user trust and loyalty, driving sustained growth and adoption within the web3 ecosystem.

Furthermore, secure storage of wallets and devices is essential for regulatory compliance and adherence to industry standards. With the increasing scrutiny from regulatory bodies and governing authorities, web3 businesses must demonstrate robust security practices to mitigate legal and compliance risks.[396] Secure storage

---

[394] Suratkar, Saurabh, Mahesh Shirole, and Sunil Bhirud. "Cryptocurrency wallet: A review." In *2020 4th international conference on computer, communication and signal processing (ICCCSP)*, pp. 1-7. IEEE, 2020.
[395] *Id.*

solutions enable businesses to implement encryption, access controls, and audit trails, thereby ensuring compliance with data protection regulations such as GDPR, CCPA, and PCI DSS. By adopting a proactive approach to security, businesses can mitigate regulatory fines, legal liabilities, and reputational damage associated with data breaches and non-compliance, fostering a culture of trust and transparency in the web3 industry.

Lastly, secure storage of wallets and devices bolsters the resilience of web3 businesses against emerging threats and cyber-attacks.[397] As cybercriminals continue to evolve their tactics and techniques, businesses must stay ahead of the curve by implementing robust security measures and proactive threat detection mechanisms. Secure storage solutions offer protection against common attack vectors such as malware, phishing, and ransomware, mitigating the risk of data loss and financial harm.[398] By investing in secure storage infrastructure and employing threat intelligence solutions, businesses can enhance their cybersecurity posture and effectively defend against evolving threats in the dynamic web3 landscape.

Using Tamper Resistant Hardware

In the rapidly evolving landscape of web3, where security and trust are paramount, tamper-resistant hardware emerges as a critical component in safeguarding digital assets and sensitive information.[399] Tamper-resistant hardware refers

---

[396] Albayati, H., Kim, S.K. and Rho, J.J., 2021. A study on the use of cryptocurrency wallets from a user experience perspective. *Human Behavior and Emerging Technologies*, *3*(5), pp.720-738

[397] Jørgensen, K.P. and Beck, R., 2022. Universal wallets. *Business & Information Systems Engineering*, pp.1-11.

[398] Albayati, *supra*.

[399] Chong, C.N., Peng, Z. and Hartel, P.H., 2003, May. Secure audit

to specialized hardware components designed to resist physical attacks and unauthorized manipulation, ensuring the integrity and confidentiality of data stored within. In the context of web3, where decentralized finance (DeFi), non-fungible tokens (NFTs), and decentralized applications (dApps) thrive, tamper-resistant hardware plays a pivotal role in protecting cryptographic keys, transaction signatures, and private data from malicious actors and cyber threats.[400]

One of the key challenges in the web3 era is the secure storage and management of cryptographic keys, which serve as the foundation for accessing and transacting digital assets on blockchain networks.[401] Tamper-resistant hardware, such as hardware security modules (HSMs) and secure enclaves, provides a secure and isolated environment for storing and processing cryptographic keys, safeguarding them from unauthorized access and tampering attempts. By leveraging advanced encryption techniques and physical security mechanisms, tamper-resistant hardware ensures that cryptographic keys remain protected even in the face of sophisticated attacks, thereby enhancing the overall security posture of web3 applications and platforms.

Moreover, tamper-resistant hardware enables secure execution environments for decentralized applications (dApps), smart contracts, and blockchain transactions.[402]

---

logging with tamper-resistant hardware. In *IFIP International Information Security Conference* (pp. 73-84). Boston, MA: Springer US.

[400] Anderson, R. and Kuhn, M., 1997, April. Low cost attacks on tamper resistant devices. In *International Workshop on Security Protocols* (pp. 125-136). Berlin, Heidelberg: Springer Berlin Heidelberg.

[401] *Id.*

By creating isolated execution environments or secure enclaves within hardware devices, developers can ensure the confidentiality, integrity, and authenticity of code execution, mitigating the risk of code tampering, injection attacks, and unauthorized access to sensitive data. This not only enhances the security and trustworthiness of dApps and smart contracts but also instills confidence among users and stakeholders in the reliability and integrity of the underlying infrastructure.

In addition to protecting cryptographic keys and executing secure transactions, tamper-resistant hardware plays a crucial role in mitigating supply chain attacks and hardware-level vulnerabilities.[403] With the increasing complexity of hardware components and the proliferation of interconnected devices in the web3 ecosystem, the risk of supply chain compromises and hardware-level exploits poses significant challenges to security and trust. Tamper-resistant hardware employs techniques such as secure boot, hardware attestation, and runtime integrity checks to detect and prevent unauthorized modifications to firmware, software, and configuration settings, thereby reducing the attack surface and strengthening the resilience of web3 infrastructure.

Tamper-resistant hardware facilitates compliance with regulatory requirements and industry standards by providing robust security controls and audit capabilities.[404] In regulated industries such as finance, healthcare, and government, where data privacy and security regulations are stringent, tamper-resistant hardware offers assurance

---

[402] Chong, *supra*.
[403] *Id.*
[404] Saggese, G.P., Romano, L., Mazzocca, N. and Mazzeo, A., 2004. A tamper resistant hardware accelerator for RSA cryptographic applications. *Journal of Systems Architecture*, *50*(12), pp.711-727.

that sensitive information is adequately protected against unauthorized access, disclosure, or manipulation. By demonstrating adherence to security best practices and regulatory guidelines, organizations operating in the web3 space can build trust with customers, partners, and regulatory authorities, fostering a secure and compliant environment for digital transactions and data exchange.[405]

Secure Disposal of Outdated or Decommissioned Hardware

The secure disposal of outdated or decommissioned hardware emerges as a critical aspect of maintaining robust security protocols. As companies upgrade their hardware infrastructure or retire legacy systems, proper disposal procedures become imperative to prevent data breaches, unauthorized access, and environmental hazards. Secure disposal involves comprehensive strategies to ensure that sensitive information stored on decommissioned hardware is permanently erased or rendered inaccessible, mitigating the risk of data leakage or compromise.

One fundamental aspect of secure disposal is data sanitization, which involves the thorough erasure or destruction of data stored on decommissioned hardware devices. Employing certified data wiping tools and techniques, such as overwriting, degaussing, or physical destruction, ensures that all traces of sensitive information, including cryptographic keys, user credentials, and proprietary data, are effectively removed from storage media.[406] By adhering to established data sanitization standards and guidelines, organizations can prevent unauthorized access to confidential data and uphold the

---

[405] Chong, *supra*.
[406] Storey, N.R., 1996. *Safety critical computer systems*. Addison-Wesley Longman Publishing Co., Inc.

integrity of their information assets throughout the disposal process.

Another key consideration in secure disposal practices is the implementation of asset tracking and inventory management systems to accurately identify and catalog decommissioned hardware assets. By maintaining a comprehensive record of hardware assets, including serial numbers, asset tags, and configuration details, organizations can track the lifecycle of each device from acquisition to disposal. This enables efficient asset retirement planning and ensures that all decommissioned hardware undergoes proper disposal procedures in accordance with security policies and regulatory requirements.

Furthermore, secure disposal encompasses environmental stewardship practices to minimize the ecological impact of electronic waste (e-waste) generated from decommissioned hardware. Responsible recycling and environmentally friendly disposal methods, such as certified e-waste recycling programs or refurbishment initiatives, help divert electronic components from landfills and promote sustainable resource management. By partnering with reputable e-waste recycling facilities or engaging in corporate sustainability initiatives, organizations can contribute to the circular economy and reduce their carbon footprint while safeguarding sensitive data during hardware disposal.

Additionally, secure disposal protocols should include physical security measures to prevent unauthorized access to decommissioned hardware assets awaiting disposal. Implementing secure storage facilities, access controls, and chain-of-custody procedures ensures that decommissioned hardware remains protected from theft, tampering, or unauthorized removal during transit or storage. By maintaining strict control over access to decommissioned hardware repositories, organizations mitigate the risk of data breaches or security incidents stemming from physical compromise of hardware assets.

Overall, adopting comprehensive and systematic approaches to secure disposal of outdated or decommissioned hardware is essential for maintaining data security, regulatory compliance, and environmental sustainability in the web3 era. By integrating data sanitization, asset tracking, environmental stewardship, and physical security measures into their disposal strategies, organizations can effectively mitigate risks associated with hardware retirement while upholding their commitment to safeguarding sensitive information and preserving the integrity of their operations.

Conclusion

Hardware security stands as a cornerstone in the realm of web3 technologies, with the secure disposal of outdated or decommissioned hardware playing a pivotal role in safeguarding sensitive data and upholding robust security protocols. By prioritizing secure storage practices for wallets and devices, organizations can mitigate the risk of unauthorized access and data breaches, ensuring the confidentiality and integrity of their digital assets. Moreover, embracing tamper-resistant hardware solutions empowers companies to fortify their defenses against physical attacks and tampering attempts, enhancing the overall resilience of their hardware infrastructure.

Furthermore, the implementation of secure disposal procedures underscores a commitment to data privacy, regulatory compliance, and environmental sustainability.

By adhering to established data sanitization standards and environmental stewardship practices, organizations can mitigate the risk of data leakage, comply with regulatory requirements, and minimize the ecological impact of electronic waste. Additionally, robust asset tracking and physical security measures contribute to maintaining accountability and integrity throughout the hardware disposal process, reducing the likelihood of security incidents or breaches stemming from mishandling of decommissioned hardware.

Ultimately, by embracing a holistic approach to hardware security and disposal, organizations can effectively manage risks, protect sensitive information, and uphold the trust and confidence of their stakeholders in an increasingly interconnected digital landscape. As the web3 ecosystem continues to evolve and expand, prioritizing hardware security remains paramount for ensuring the resilience, integrity, and sustainability of digital infrastructure in the face of emerging threats and challenges. Through proactive measures and strategic investments in hardware security, organizations can navigate the complexities of the digital age with confidence and resilience, safeguarding their assets and preserving the trust of their customers and partners.

Chapter 18: Data Center Security

"Security is mostly superstition. It does not exist in nature, nor do the children of men as a whole experience it. Avoiding danger is no safer in the long run than outright exposure. Life is either a daring adventure or nothing."

Hellen Keller[407]

Data center security emerges as a critical imperative for organizations leveraging web3 technologies. As the nucleus of digital operations, data centers house vast repositories of sensitive information, including server farms, necessitating robust security measures to safeguard against unauthorized access, environmental hazards, and unforeseen disasters. The importance of data center security extends beyond mere protection of assets; it encompasses the preservation of data integrity, continuity of operations, and compliance with regulatory frameworks governing data privacy and security.

Physical access control stands as a foundational pillar of data center security, dictating who can enter and interact with the infrastructure housing critical digital assets. Implementing stringent access control protocols, such as biometric authentication, surveillance systems, and restricted entry points, mitigates the risk of unauthorized intrusion and physical tampering, bolstering the overall resilience of the data center environment. By limiting access to authorized personnel and employing robust authentication mechanisms, organizations can fortify their defenses against insider threats and external attacks, safeguarding the integrity and confidentiality of their data.

---

[407] (2024). "Security Quotes, Hellen Keller." Brainy Quote at https://www.brainyquote.com/topics/security-quotes.

Moreover, environmental controls play a pivotal role in data center security, addressing vulnerabilities posed by temperature fluctuations, humidity levels, and other environmental factors that can compromise hardware performance and data integrity. Deploying state-of-the-art environmental monitoring systems and climate control technologies enables organizations to maintain optimal operating conditions within data centers, minimizing the risk of equipment failure, downtime, and data loss due to environmental hazards. Additionally, implementing redundant power sources, fire suppression systems, and physical security measures further enhances the resilience of data center infrastructure against unforeseen contingencies and natural disasters.

In the event of a disruptive incident or catastrophic event, having a comprehensive disaster recovery plan in place is essential for mitigating the impact on data center operations and minimizing downtime. Establishing robust backup and recovery procedures, offsite data replication, and failover mechanisms ensures continuity of operations and rapid restoration of services in the face of adverse events, such as power outages, hardware failures, or cyberattacks. By proactively planning for contingencies and rehearsing response strategies, organizations can effectively mitigate the risk of data loss, financial repercussions, and reputational damage associated with unplanned disruptions.

Furthermore, the secure disposal of documents and sensitive materials is paramount for maintaining data integrity, confidentiality, and regulatory compliance throughout the data lifecycle. Implementing secure document destruction protocols, such as shredding and incineration, ensures that sensitive information is irreversibly rendered unreadable and inaccessible, mitigating the risk of unauthorized disclosure or misuse. By adhering to industry best practices and regulatory guidelines for data disposal, organizations can demonstrate accountability, transparency, and commitment to protecting the privacy and security of sensitive information.

Data center security encompasses a multifaceted approach to safeguarding digital assets, encompassing physical access controls, environmental monitoring, disaster recovery planning, and secure data disposal practices. By prioritizing robust security measures and proactive risk management strategies, organizations can fortify their defenses against evolving threats, ensure the continuity of operations, and uphold the trust and confidence of stakeholders in an increasingly interconnected digital ecosystem. As the backbone of digital infrastructure, data centers play a pivotal role in driving innovation, powering critical business processes, and enabling the seamless delivery of services in the digital age.

Physical Access to the System

Controlling access to sensitive data is paramount for any company, regardless of its size or industry. In today's digital landscape, where data has become a valuable asset and a prime target for cyber threats, safeguarding sensitive information is essential to protect intellectual property, maintain customer trust, and comply with regulatory requirements.[408] By implementing robust access control mechanisms, organizations can mitigate the risk of unauthorized disclosure, misuse, or tampering of critical data, thereby preserving its confidentiality, integrity, and availability.

---

[408] Jazdi, N., 2014, May. Cyber physical systems in the context of Industry 4.0. In *2014 IEEE international conference on automation, quality and testing, robotics* (pp. 1-4). IEEE.

Companies have both the right and the responsibility to protect their data from unauthorized access or exploitation. As custodians of sensitive information, organizations are entrusted with safeguarding the privacy and confidentiality of data belonging to their employees, customers, and stakeholders.[409] Failing to adequately secure this data not only exposes the company to legal and financial liabilities but also undermines its reputation and erodes trust among stakeholders.[410] By proactively managing access permissions and enforcing security policies, organizations can demonstrate their commitment to data protection and earn the confidence of their constituents.

Furthermore, there are instances where the convenience of easy access to data can inadvertently pose security risks, even for seemingly innocuous information. In today's interconnected and fast-paced business environment, employees often require access to a wide range of data to perform their duties effectively. However, granting unrestricted access to sensitive information increases the likelihood of data breaches, insider threats, and inadvertent data leaks.[411] Therefore, striking the right balance between accessibility and security is crucial, necessitating the implementation of granular access controls and least privilege principles to limit access to only those who genuinely need it.

Moreover, the proliferation of remote work and mobile devices has further heightened the importance of controlling access to sensitive data.[412] With employees

---

[409] *Id.*
[410] Hu, V.C., Ferraiolo, D. and Kuhn, D.R., 2006. *Assessment of access control systems* (pp. 09-29). Gaithersburg, MD: US Department of Commerce, National Institute of Standards and Technology
[411] *Id.*
[412] Jazdı, *supra.*

accessing corporate networks and data from various locations and devices, ensuring secure and authenticated access becomes increasingly challenging. Adopting robust authentication methods, such as multi-factor authentication (MFA) and role-based access control (RBAC), can help mitigate the risks associated with remote access and strengthen the overall security posture of the organization.

Controlling access to sensitive data is a fundamental aspect of data security strategy for any organization.[413] By implementing effective access control mechanisms, companies can protect their valuable assets, uphold regulatory compliance, and mitigate the risks posed by internal and external threats. Additionally, by prioritizing security without compromising usability, organizations can strike a balance between accessibility and protection, thereby fostering a culture of data security and resilience across the enterprise.

Environmental Controls (and Fire Security)

Environmental controls play a critical role in ensuring the reliability and security of data centers, but they can also introduce vulnerabilities that malicious actors may exploit to gain unauthorized access to sensitive systems and data. One of the primary concerns is the risk of physical intrusion facilitated by compromised environmental systems, such as fire management and HVAC (heating, ventilation, and air conditioning) systems.[414] While these systems are designed to maintain optimal operating

---

[413] Atat, R., Liu, L., Wu, J., Li, G., Ye, C. and Yang, Y., 2018. Big data meet cyber-physical systems: A panoramic survey. *IEEE Access*, *6*, pp.73603-73636.
[414] Naedele, M., 2007, January. Addressing IT security for critical control systems. In *2007 40th Annual Hawaii International Conference on System Sciences (HICSS'07)* (pp. 115-115). IEEE..

conditions within the data center, they can inadvertently provide avenues for attackers to bypass traditional security measures.

For example, attackers may exploit vulnerabilities in fire suppression systems to disable or tamper with critical infrastructure, such as fire alarms or sprinkler systems, to create distractions or disrupt operations while they carry out their malicious activities.[415] Similarly, HVAC systems, which regulate temperature and humidity levels in the data center, can be targeted to gain unauthorized access by compromising access controls or exploiting vulnerabilities in the network infrastructure connected to these systems. By infiltrating through less secure routes, such as air vents or ducts, attackers can bypass physical barriers and gain proximity to sensitive equipment and data.[416]

Moreover, environmental controls, such as temperature and humidity sensors, may inadvertently leak information or provide insights into the operational status of the data center, which could be leveraged by attackers to plan and execute their attacks more effectively.[417] For instance, fluctuations in temperature or humidity levels could indicate periods of reduced activity or maintenance windows, allowing attackers to time their intrusion attempts for maximum impact. Additionally, environmental monitoring systems that transmit data over insecure channels or lack proper encryption may expose sensitive information to interception or tampering by malicious actors.

---

[415] Krotofil, M. and Gollmann, D., 2013, July. Industrial control systems security: What is happening?. In *2013 11th IEEE International Conference on Industrial Informatics (INDIN)* (pp. 670-675). IEEE.
[416] Benantar, M., 2005. *Access control systems: security, identity management and trust models*. Springer Science & Business Media.
[417] Naedele, *supra*.

Another concern is the potential exploitation of accessibility systems, such as electronic locks or biometric scanners, which are used to control physical access to the data center facilities. If these systems are not properly secured or integrated with robust authentication mechanisms, attackers could compromise them to gain unauthorized entry into restricted areas.[418] For example, attackers may attempt to clone access credentials or exploit vulnerabilities in access control systems to bypass authentication mechanisms and gain physical access to critical infrastructure.

To mitigate these risks, organizations must implement comprehensive security measures that encompass not only traditional cybersecurity controls but also physical security protocols and environmental safeguards. This includes conducting regular assessments of environmental systems for vulnerabilities, implementing intrusion detection and monitoring mechanisms to detect anomalous activities, and establishing stringent access controls and authentication measures to restrict access to authorized personnel only. Additionally, organizations should prioritize employee training and awareness programs to educate staff about the importance of physical security and vigilance in detecting and reporting suspicious activities within the data center environment.[419] By adopting a holistic approach to security that addresses both cyber and physical threats, organizations can enhance their resilience and protect their data assets from evolving threats and vulnerabilities.

---

[418] Krotofil, *supra*.
[419] *Id.*

Disaster Recovery Plan

Disaster recovery plans are paramount for organizations to ensure the security and integrity of sensitive data in the aftermath of emergencies or unforeseen events. These plans outline systematic procedures and protocols to be followed in the event of a disaster, such as natural disasters, cyberattacks, or system failures, with the primary goal of minimizing downtime and data loss while restoring operations to normalcy as swiftly as possible.[420]

First, a well-defined disaster recovery plan provides a structured framework for responding to emergencies, enabling organizations to quickly assess the impact of the disaster on their data center infrastructure and prioritize recovery efforts accordingly. By establishing clear roles, responsibilities, and communication channels, organizations can streamline decision-making processes and coordinate response efforts more effectively, reducing the risk of confusion or delays during critical moments.[421] This proactive approach helps minimize the potential disruption to business operations and mitigate the financial and reputational impacts associated with prolonged downtime or data loss.

Furthermore, disaster recovery plans play a crucial role in safeguarding sensitive data by implementing robust backup and recovery mechanisms to ensure data availability and integrity. By regularly backing up critical data and storing redundant copies in geographically diverse locations or

---

[420] Maiwald, E. and Sieglein, W., 2002. *Security planning & disaster recovery*. McGraw-Hill/Osborne.

[421] El-Temtamy, O., Majdalawieh, M. and Pumphrey, L., 2016. Assessing IT disaster recovery plans: The case of publicly listed firms on Abu Dhabi/UAE security exchange. *Information & Computer Security*, *24*(5), pp.514-533.

cloud-based environments, organizations can minimize the risk of permanent data loss and facilitate timely data restoration in the event of a disaster.[422] Additionally, disaster recovery plans typically include procedures for testing and validating backup systems and recovery processes to identify and address any gaps or vulnerabilities proactively. This iterative approach ensures the reliability and effectiveness of the recovery strategy, enhancing the organization's overall resilience to unforeseen threats and disruptions.

Disposal of Documents (and document type)

Proper disposal of documents is critical for maintaining data security and preventing unauthorized access to sensitive information.[423] Even with advancements in document shredding technologies like cross-cut shredders, there are sophisticated computer algorithms capable of reconstructing shredded documents, posing a significant risk to data confidentiality if not handled correctly. While cross-cut shredding may render documents into tiny fragments, the potential for reconstruction remains, particularly when dealing with patterns or structured layouts on the paper, which can aid in piecing together the shredded pieces.

The proliferation of digital scanning and imaging technologies has further exacerbated the risk of document reconstruction, as digital images of shredded documents can be analyzed and digitally reconstructed using specialized software tools.[424] This underscores the

---

[422] *Id.*
[423] Mansilla, G. and Verde, M., 2015. Disposal of documents: a proposal for libraries. *Investigación Bibliotecológica: Archivonomía, Bibliotecología e Información*, *29*(67), pp.91-111.
[424] Wewers, T. and Wargitsch, C., 1998, January. Four dimensions of

importance of adopting additional measures beyond traditional shredding methods to ensure the complete and irretrievable destruction of sensitive documents, especially those containing confidential or proprietary information.

To mitigate the risk of document reconstruction and data exposure, organizations must implement comprehensive document disposal policies and procedures that address the specific challenges associated with document destruction.[425] This may involve employing more secure shredding techniques, such as micro-cut shredding, which reduces documents to even smaller particles, making reconstruction significantly more difficult and time-consuming for potential adversaries.

Organizations should consider implementing protocols for the secure disposal of electronic media and digital storage devices, as these can also contain sensitive information that may be targeted by malicious actors if not properly disposed of.[426] This may include degaussing or physically destroying storage devices to ensure data irreversibility and prevent data recovery attempts by unauthorized parties.

Ultimately, effective document disposal practices are essential components of a holistic data security strategy, complementing other security measures such as encryption, access controls, and data loss prevention.[427] By proactively

interorganizational, document-oriented workflow: a case study of the approval of hazardous-waste disposal. In *Proceedings of the Thirty-First Hawaii International Conference on System Sciences* (Vol. 4, pp. 332-341). IEEE.

425 *Id.*

426 Wells, F.L., Winston, G.B. and Beers, H.P., 1944. Historical development of the records disposal policy of the Federal Government prior to 1934. *The American Archivist*, *7*(3), pp.181-201.

427 Wewers & Wargitsch, *supra.*

addressing the risks associated with document disposal and adopting best practices for data destruction, organizations can enhance their overall security posture and safeguard sensitive information from unauthorized access or exploitation.

Conclusion

Data center security is paramount for safeguarding sensitive information and maintaining the integrity of organizational data assets. The multifaceted nature of data center security encompasses various aspects, including controlling physical access to the facility, implementing robust environmental controls, devising comprehensive disaster recovery plans, and ensuring proper disposal of documents. By addressing these critical components, organizations can significantly mitigate the risk of unauthorized access, data breaches, and potential disruptions to business operations.

Effective access controls play a pivotal role in preventing unauthorized individuals from gaining entry to data center facilities and compromising sensitive information. Access management protocols, such as biometric authentication, access cards, and surveillance systems, help restrict access to authorized personnel only, bolstering overall security posture. Additionally, implementing stringent environmental controls, such as fire suppression systems and temperature regulation mechanisms, mitigates the risk of environmental hazards and protects hardware infrastructure from damage or malfunction.

Furthermore, robust disaster recovery plans are essential for minimizing the impact of unforeseen events, such as natural disasters or cyberattacks, on data center operations and ensuring timely recovery of critical systems and data.

These plans outline procedures for data backup, replication, and restoration, enabling organizations to recover swiftly and resume normal business activities in the event of a disruption. Additionally, proper disposal of documents, both physical and electronic, is crucial for preventing data leakage and ensuring compliance with regulatory requirements pertaining to data privacy and security.

By prioritizing data center security and implementing comprehensive security measures tailored to their specific needs, organizations can effectively safeguard their data assets and uphold the trust and confidence of their stakeholders. Data security is an ongoing process that requires continuous vigilance, proactive risk management, and adherence to industry best practices. By adopting a proactive and holistic approach to data center security, organizations can enhance their resilience to emerging threats and ensure the confidentiality, integrity, and availability of their critical data resources.

Chapter 19: Employee Security

"Ninety percent of people support background checks. Which means even people who can't pass a background check support background checks."

Bill Maher[428]

Employee security is a cornerstone of comprehensive cybersecurity measures within organizations, as employees represent both a critical asset and a potential vulnerability in safeguarding sensitive information and mitigating cyber risks. In today's interconnected digital world, where cyber threats continue to evolve and proliferate, ensuring the security posture of employees is paramount to protecting organizational assets, maintaining regulatory compliance, and preserving brand reputation. This chapter delves into the multifaceted aspects of employee security, emphasizing key strategies and best practices for enhancing employee awareness, mitigating insider threats, and fostering a culture of security within the workplace.

Background checks serve as an essential first line of defense in vetting prospective employees and identifying potential security risks before they are onboarded. Conducting thorough background checks enables organizations to assess an individual's criminal history, employment background, and other relevant credentials, helping to mitigate the risk of insider threats and ensure the integrity of the workforce. By implementing robust background screening processes, organizations can make informed hiring decisions and mitigate the risk of insider threats posed by malicious actors seeking employment within the organization.

---

[428] (2024). "Background Checks, Bill Maher." AZ Quotes at https://www.azquotes.com/quotes/topics/background-checks.html.

Security awareness training plays a pivotal role in empowering employees with the knowledge and skills needed to recognize and respond effectively to various cyber threats and social engineering tactics. Through comprehensive security awareness programs, employees can learn about the latest cybersecurity threats, phishing scams, and best practices for safeguarding sensitive information. By educating employees on security protocols, policies, and procedures, organizations can strengthen their defense against cyber threats and cultivate a security-conscious culture where employees are vigilant and proactive in identifying and reporting potential security incidents.

Policy training is instrumental in ensuring that employees understand and adhere to organizational policies, procedures, and guidelines governing information security and data protection. By providing regular policy training sessions, organizations can reinforce compliance with regulatory requirements, industry standards, and internal policies, thereby reducing the risk of policy violations and data breaches. Additionally, policy training helps employees understand their roles and responsibilities in safeguarding sensitive information and reinforces the importance of adhering to security protocols and best practices.

Limiting access to restricted materials and sensitive data is essential for minimizing the risk of insider threats and unauthorized access to confidential information. By implementing role-based access controls and least privilege principles, organizations can restrict access to sensitive data only to employees who require it to perform their job functions. This proactive approach helps mitigate the risk of insider threats and ensures that sensitive information remains protected from unauthorized access or misuse.

Creating a culture of "see something, say something" encourages employees to actively participate in identifying and reporting security incidents, suspicious activities, or potential vulnerabilities. By fostering open communication channels and promoting a culture of collaboration and accountability, organizations can empower employees to play an active role in maintaining a secure work environment. Encouraging employees to report security concerns promptly enables organizations to address potential threats in a timely manner and mitigate the risk of security incidents or data breaches.

Employee security encompasses a range of strategies and initiatives aimed at mitigating insider threats, enhancing security awareness, and fostering a culture of vigilance and accountability within the workplace. By prioritizing employee security and implementing comprehensive security measures, organizations can strengthen their overall security posture, mitigate the risk of insider threats, and protect sensitive information from unauthorized access or misuse.

Background Checks

Background checks serve as a critical component of pre-employment screening processes, enabling organizations to assess the trustworthiness and suitability of prospective employees before they are granted access to sensitive information or entrusted with key responsibilities.[429] By conducting thorough background checks, organizations can verify the credentials, employment history, and criminal

---

[429] Holzer, H.J., Raphael, S. and Stoll, M.A., 2001. Will Employers Hire Ex-Offenders? Employer Checks, Background Checks, and Their. *Info:*.

background of applicants, helping to identify any red flags or potential security risks early in the hiring process.

Mitigating the risk of insider threats is paramount in safeguarding organizational assets and maintaining the integrity of operations.[430] Background checks provide valuable insights into an individual's past behavior and affiliations, helping organizations identify candidates who may pose a security risk due to previous criminal activities, associations with malicious entities, or other concerning factors. By screening candidates rigorously, organizations can mitigate the risk of insider threats and protect against internal vulnerabilities that could compromise security.[431]

From a compliance standpoint, background checks are often mandated by regulatory requirements or industry standards, particularly in sectors handling sensitive or regulated data.[432] By ensuring compliance with background screening regulations, organizations demonstrate their commitment to due diligence and risk management, thereby reducing the likelihood of legal and regulatory repercussions associated with negligent hiring practices or security breaches stemming from insider threats.

Beyond risk mitigation, background checks also contribute to fostering a culture of trust and accountability within the organization.[433] By demonstrating a commitment to hiring

---

[430] Levashina, J. and Campion, M.A., 2009. Expected practices in background checking: Review of the human resource management literature. *Employee Responsibilities and Rights Journal, 21*, pp.231-249.
[431] *Id.*
[432] Holzer, Rapheal & Stole, *supra.*
[433] Levashina, J. and Campion, M.A., 2009. Expected practices in background checking: Review of the human resource management literature. *Employee Responsibilities and Rights Journal, 21*, pp.231-249.

candidates with integrity and credibility, organizations instill confidence among employees and stakeholders, reinforcing the importance of ethical conduct and upholding the organization's reputation as a trustworthy and responsible entity in the eyes of customers, partners, and the broader community.

Security Awareness Training

Security awareness training plays a pivotal role in equipping employees with the knowledge, skills, and vigilance needed to identify and respond effectively to various cyber threats, phishing attacks, and social engineering tactics. By educating employees about potential risks and security best practices, organizations empower them to become active participants in safeguarding sensitive information and protecting against cyber threats.

In today's dynamic and evolving threat landscape, where cybercriminals constantly devise new tactics to exploit vulnerabilities, ongoing security awareness training is essential for keeping employees informed and prepared to recognize emerging threats.[434] By staying abreast of the latest cybersecurity trends and threat vectors, employees can remain vigilant and proactive in identifying potential security incidents, thereby enhancing the organization's overall resilience to cyber-attacks.

Beyond technical safeguards and security controls, human factors remain one of the most significant vulnerabilities in organizational security. Security awareness training helps

---

[434] Eminağaoğlu, M., Uçar, E. and Eren, Ş., 2009. The positive outcomes of information security awareness training in companies–A case study. *information security technical report*, *14*(4), pp.223-229.

address this vulnerability by raising awareness about common pitfalls, such as clicking on suspicious links, sharing sensitive information inadvertently, or falling victim to social engineering scams.[435] By instilling a culture of security awareness, organizations empower employees to make informed decisions and adopt security-conscious behaviors in their day-to-day activities.

Effective security awareness training programs go beyond mere compliance requirements to foster a culture of security consciousness and personal responsibility among employees. By promoting a shared understanding of security risks and responsibilities across the organization, security awareness training encourages employees to take ownership of their role in safeguarding sensitive information and protecting the organization's assets. This collective commitment to security not only strengthens the organization's defenses against cyber threats but also cultivates a culture of trust, collaboration, and resilience in the face of evolving security challenges.

Policy and Procedure Training (how it is different from awareness training)

Policy and procedure training is instrumental in ensuring that employees understand and adhere to the security protocols and guidelines established by the organization to safeguard sensitive information and mitigate security risks.[436]

---

[435] Shaw, R.S., Chen, C.C., Harris, A.L. and Huang, H.J., 2009. The impact of information richness on information security awareness training effectiveness. *Computers & Education*, *52*(1), pp.92-100.
[436] Laksonen Jr, R.P. and Gasiewicz, N.K., 2015. Implementing a Program for Ultrasound-Guided Peripheral Venous Access: Training, Policy and Procedure Development, Protocol Use, Competency, and

Through comprehensive training programs, employees are educated on the company's security policies, procedures, and best practices, covering aspects such as data handling, access control, incident response, and compliance requirements. By familiarizing employees with these policies and instilling a culture of compliance, organizations can enhance their overall security posture and minimize the likelihood of security breaches resulting from human error or negligence.

Effective policy and procedure training should encompass various topics relevant to web3 company security, including data encryption, password management, secure communication protocols, and regulatory compliance. Employees should receive clear instructions on how to handle sensitive data securely, the importance of using strong authentication methods, and the procedures for reporting security incidents or suspicious activities.[437] By providing employees with the knowledge and resources they need to adhere to security policies effectively, organizations can foster a culture of accountability and vigilance across all levels of the workforce.

In addition to training on specific security policies and procedures, employees should also receive regular updates and refresher courses to stay informed about evolving security threats and emerging best practices. As the threat landscape continues to evolve, continuous education and awareness are essential to ensure that employees remain vigilant and adaptable in the face of new challenges. By investing in ongoing training initiatives, organizations

---

Skill Tracking. *The Nursing clinics of North America*, *50*(4), pp.771-785.

[437] Mickle, A.M., 2001. *Athletic training risk management: Developing policy and procedure from case law analysis*. New Mexico State University.

demonstrate their commitment to maintaining a proactive approach to security and equipping employees with the skills and knowledge needed to protect against evolving threats effectively.

Web3 companies can further enhance their security posture by implementing a zoning model that compartmentalizes access to sensitive information based on the principle of least privilege. By structuring their business operations into distinct security zones, such as public, internal, and restricted zones, organizations can enforce granular access controls and restrict employees' access to sensitive data only to the extent necessary for their roles and responsibilities.[438] This zoning approach helps minimize the risk of unauthorized access and data exposure, reducing the potential impact of security breaches and enhancing overall data protection measures. Additionally, by delineating clear boundaries between different security zones and implementing robust access controls, organizations can better contain and mitigate the impact of security incidents or breaches, thereby safeguarding their assets and maintaining the confidentiality, integrity, and availability of critical information.

See Something, Say Something

Encouraging a culture of "see something, say something" is essential for fostering a proactive approach to security within web3 companies. This ethos emphasizes the importance of individual vigilance and collective responsibility in identifying and reporting potential security threats or suspicious activities. By empowering employees to speak up about any security concerns they observe, organizations can create a collaborative environment where

---

[438] *Id.*

everyone plays a role in safeguarding sensitive information and mitigating security risks. This proactive stance helps detect and address security incidents in their early stages, minimizing their impact and preventing more significant breaches or disruptions.

To instill a culture of "see something, say something," organizations must prioritize open communication, transparency, and trust between employees and management. Employees should feel comfortable raising security-related concerns without fear of retribution or reprisal, knowing that their observations will be taken seriously and addressed promptly. By fostering an environment where communication channels are accessible, responsive, and supportive, organizations can create a sense of collective responsibility for security and encourage employees to actively participate in protecting the company's assets and reputation.

Effective communication and awareness initiatives are key to promoting the "see something, say something" culture throughout the organization. Regular security awareness training sessions, workshops, and communication campaigns can educate employees about common security threats, warning signs of suspicious activity, and the importance of reporting incidents promptly. Providing clear guidelines and protocols for reporting security concerns, as well as establishing designated points of contact or reporting channels, helps streamline the reporting process and ensure that relevant information reaches the appropriate stakeholders in a timely manner.

Recognizing and rewarding employees for their contributions to security awareness and incident reporting can further reinforce the desired behavior and incentivize active participation in the "see something, say something" initiative. By acknowledging and appreciating employees' vigilance and proactive efforts to maintain security, organizations reinforce the value of individual contributions to the collective security posture. This positive reinforcement encourages continued engagement and helps embed the "see something, say something" mindset as a fundamental aspect of the organization's culture and values.

Beyond internal initiatives, web3 companies can also leverage external resources and partnerships to enhance their security posture and promote a culture of vigilance. Collaborating with industry peers, participating in information-sharing networks, and staying informed about emerging threats and best practices can provide valuable insights and support for strengthening security awareness and incident response capabilities. By actively engaging with the broader security community and leveraging external expertise and resources, organizations can proactively identify and address security challenges, adapt to evolving threats, and continuously improve their overall security posture.

Conclusion

Employee security stands as a foundational pillar in the defense against cyber threats within web3 companies. Through robust background checks, organizations can mitigate risks associated with insider threats and ensure that individuals with access to sensitive information possess the necessary trustworthiness and integrity. Moreover, comprehensive security awareness training equips employees with the knowledge and skills to recognize and respond to potential security incidents, bolstering the organization's overall resilience to cyber threats.

Policy and procedure training further fortifies employee security by establishing clear guidelines and protocols for handling sensitive information and accessing restricted materials. By structuring business operations into zones and limiting access based on job roles and responsibilities, organizations can enforce the principle of least privilege and minimize the risk of unauthorized access or data breaches.

A culture of "see something, say something" serves as a powerful deterrent against security threats by empowering employees to actively participate in identifying and reporting suspicious activities. By fostering open communication, transparency, and trust, organizations can create an environment where security concerns are promptly addressed, and proactive measures are taken to mitigate potential risks.

In essence, employee security is not merely a compliance requirement but a strategic imperative for web3 companies seeking to safeguard their assets, reputation, and long-term viability in an increasingly complex threat landscape. By investing in the development of a resilient security culture and providing employees with the tools, training, and support they need to fulfill their role in protecting the organization, companies can effectively mitigate security risks and build a strong defense against cyber threats.

Chapter 20: Supply Chain Security

"He who controls the spice controls the universe."

Frank Herbert, DUNE[439]

Supply chain management plays a pivotal role in ensuring the security and integrity of devices and facilities within web3 companies. Conducting thorough vendor risk assessments is essential to identify potential vulnerabilities and mitigate risks associated with third-party suppliers. By evaluating factors such as the vendor's security practices, regulatory compliance, and track record of security incidents, organizations can make informed decisions about their partnerships and proactively address any security concerns.

Maintaining a secure delivery system or facility is paramount to safeguarding the integrity of products and materials throughout the supply chain. Implementing stringent access controls, surveillance systems, and physical security measures helps prevent unauthorized access and tampering, reducing the risk of security breaches or compromises during transit or storage. By establishing secure delivery protocols and ensuring the integrity of receiving facilities, organizations can minimize the likelihood of supply chain-related security incidents.

Ensuring a secure chain of custody is essential for tracking the movement and handling of products, components, or sensitive information throughout the supply chain. Implementing robust documentation and verification procedures enables organizations to maintain visibility and accountability at each stage of the supply chain, reducing

---

[439] (2024). "Supply Chain Quotes, Frank Herbert." Good Reads at https://www.goodreads.com/quotes/tag/supply-chain.

the risk of loss, theft, or unauthorized access. By implementing digital tracking technologies and secure authentication mechanisms, companies can enhance the traceability and integrity of their supply chain operations.

Contractual agreements with vendors should include comprehensive provisions related to security requirements, responsibilities, and accountability. By establishing clear expectations and obligations regarding data protection, confidentiality, and compliance with security standards, organizations can mitigate legal and financial risks associated with supply chain security breaches. Moreover, including provisions for regular security audits, incident response protocols, and indemnification clauses can further strengthen contractual agreements and enhance the organization's ability to respond effectively to security incidents or breaches.

Effective supply chain management is essential for ensuring the security, resilience, and trustworthiness of devices and facilities within web3 companies. By conducting rigorous vendor risk assessments, maintaining secure delivery systems, implementing robust chain of custody procedures, and establishing airtight contractual agreements, organizations can mitigate supply chain-related security risks and safeguard their operations against potential threats and vulnerabilities.

Vendor Risk Assessment

Having a robust vendor assessment protocol is critical for web3 companies to ensure the security and integrity of their supply chain.[440] First and foremost, such protocols enable

---

[440] Sunderkrishnan, L., 2016. Vendor Risk Assessment. *EDPACS*, *54*(4), pp.19-26.

organizations to evaluate potential vendors thoroughly before engaging in business partnerships. By conducting comprehensive assessments, companies can assess the vendor's reliability, trustworthiness, and adherence to security standards.[441] This initial screening process allows organizations to identify vendors that align with their security requirements and mitigate the risk of partnering with unreliable or non-compliant suppliers.

Furthermore, robust vendor assessment protocols enable organizations to identify and address potential security risks and vulnerabilities within their supply chain. By evaluating factors such as the vendor's security practices, data protection measures, and regulatory compliance, companies can identify areas of concern and implement appropriate risk mitigation strategies. This proactive approach helps organizations prevent security breaches, data leaks, and other supply chain-related incidents that could compromise their operations and reputation.

Vendor assessment protocols help organizations establish clear expectations and requirements for their suppliers regarding security standards and compliance. By communicating these expectations upfront and incorporating them into vendor contracts and agreements, companies can ensure that vendors understand their responsibilities and commitments regarding security.[442] This transparency and accountability promote a culture of security throughout the supply chain and facilitate effective collaboration between organizations and their vendors.

---

441 *Id.*

442 Wu, D. and Olson, D.L., 2008. Supply chain risk, simulation, and vendor selection. *International journal of production economics*, *114*(2), pp.646-655.

Additionally, robust vendor assessment protocols provide a mechanism for ongoing monitoring and evaluation of vendor performance and security posture. By regularly reviewing and updating vendor assessments, organizations can track changes in the vendor's security practices, address emerging risks, and ensure ongoing compliance with security standards.[443] This continuous monitoring helps organizations stay vigilant and responsive to evolving threats and vulnerabilities in the supply chain, thereby enhancing overall resilience and risk management capabilities.

Having a robust vendor assessment protocol is essential for web3 companies to mitigate security risks, ensure compliance with security standards, and maintain the integrity of their supply chain. By conducting thorough assessments, addressing potential vulnerabilities, establishing clear expectations, and monitoring vendor performance, organizations can strengthen their supply chain security and safeguard their operations against external threats and disruptions.

Secure Delivery Facility and Chain of Custody

A secure facility for hardware deliveries is paramount to safeguarding the integrity of the supply chain and protecting sensitive equipment from theft, tampering, or unauthorized access. When hardware is delivered to a secure facility, it reduces the risk of interception by malicious actors during transit.[444] By establishing

---

[443] Briggs, P., 1994. Vendor assessment for partners in supply. *European Journal of Purchasing & Supply Management*, *1*(1), pp.49-59.
[444] Nicholson, A.J., Chawathe, Y., Chen, M.Y., Noble, B.D. and Wetherall, D., 2006, June. Improved access point selection. In *Proceedings of the 4th international conference on Mobile systems,*

designated delivery points equipped with security measures such as surveillance cameras, access controls, and trained personnel, companies can ensure that deliveries are received securely and remain under constant supervision until they are properly accounted for and processed.

Furthermore, a secure delivery facility provides a controlled environment where incoming hardware can be inspected, verified, and documented according to established protocols. This enables organizations to verify the authenticity and integrity of delivered items, ensuring that they have not been tampered with or compromised in transit. By meticulously documenting the receipt of hardware shipments, including serial numbers, packaging details, and other relevant information, companies can establish a clear chain of custody and traceability, which is essential for accountability and risk management.

In situations where a dedicated delivery facility is not feasible, having a secure location within the office for receiving deliveries becomes essential. This designated area should be equipped with security features such as secure access controls, surveillance cameras, and restricted entry to authorized personnel only.[445] By designating a secure location for deliveries, companies can minimize the risk of unauthorized access to incoming hardware and mitigate the potential for theft, tampering, or loss.

In addition to establishing a secure facility or location for hardware deliveries, maintaining accurate chain of custody documents is crucial for ensuring accountability and

---

applications and services (pp. 233-245).
[445] Vasudevan, S., Papagiannaki, K., Diot, C., Kurose, J. and Towsley, D., 2005, October. Facilitating access point selection in IEEE 802.11 wireless networks. In *Proceedings of the 5th ACM SIGCOMM Conference on Internet Measurement* (pp. 26-26).

transparency throughout the delivery process. Chain of custody documents provide a detailed record of the journey of hardware items from the point of origin to their final destination, documenting each stage of handling and custody.[446] By documenting the receipt, transfer, and storage of hardware shipments, companies can establish a clear audit trail and accountability for each item, facilitating effective tracking and monitoring of the supply chain and reducing the risk of loss or unauthorized access.

Contractual Agreements with Vendors

Robust contracts with vendors, encompassing comprehensive security protocols, is essential for mitigating risks, ensuring accountability, and safeguarding the integrity of the supply chain. First of all, a strong contract outlines the expectations and responsibilities of both parties regarding security measures, data protection, and compliance with industry standards and regulations.[447] By clearly defining these terms, companies can hold vendors accountable for adhering to agreed-upon security protocols and standards, reducing the likelihood of security breaches or data compromises.

A well-crafted contract with vendors establishes clear guidelines for risk management and liability allocation in the event of security incidents or breaches. This includes provisions for indemnification, specifying the extent of liability each party assumes for breaches or damages resulting from security lapses.[448] By delineating these terms

[446] Lovell, K. and Richards, D., 2000. Multiple access points and levels of entry (MAPLE): ensuring choice, accessibility and equity for CBT services. *Behavioural and Cognitive Psychotherapy*, *28*(4), pp.379-391.
[447] Platz, L.A. and Temponi, C., 2007. Defining the most desirable outsourcing contract between customer and vendor. *Management decision*, *45*(10), pp.1656-1666,

upfront, companies can minimize legal disputes and ensure that vendors bear appropriate responsibility for any security-related incidents that may occur during the course of the business relationship.

A strong contract enables companies to enforce security requirements and standards across the entire supply chain, including subcontractors or third-party vendors utilized by primary suppliers.[449] By stipulating stringent security measures and compliance requirements in vendor contracts, companies can exert greater control over the security posture of their suppliers and mitigate the risks posed by potential vulnerabilities or weaknesses in the supply chain ecosystem.

Contracts with vendors should include clauses addressing data privacy, intellectual property rights, confidentiality agreements, and termination clauses. These provisions help protect sensitive information, proprietary data, and trade secrets from unauthorized access, disclosure, or misuse by vendors or third parties.[450] By incorporating these safeguards into vendor contracts, companies can safeguard their intellectual property and proprietary information while fostering trust and confidence in their business relationships.

The reality is that strong contracts with vendors serve as a crucial tool for establishing mutual trust, transparency, and

---

[448] Sainathan, A. and Groenevelt, H., 2019. Vendor managed inventory contracts–coordinating the supply chain while looking from the vendor's perspective. *European Journal of Operational Research*, *272*(1), pp.249-260.
[449] *Id.*
[450] Chakraborty, S. and Philip, T., 1996. Vendor development strategies. *International Journal of Operations & Production Management*, *16*(10), pp.54-66.

accountability in business partnerships. By clearly defining security expectations, responsibilities, and liabilities, companies can mitigate risks, protect sensitive data, and uphold the integrity of their supply chain operations, ultimately enhancing resilience and enabling sustainable business growth.

Conclusion

Supply chain management plays a pivotal role in ensuring the security and integrity of web3 companies' operations. By implementing robust vendor assessment protocols, companies can assess the security posture of their suppliers, identify potential risks, and mitigate vulnerabilities in the supply chain ecosystem. Securing delivery facilities or designated areas within the office for hardware deliveries is essential for preventing unauthorized access, tampering, or theft of sensitive equipment during transit.

Maintaining accurate chain of custody documents for all incoming shipments helps track the movement of hardware components, ensuring accountability and transparency throughout the supply chain. These documents provide a clear record of who handled the equipment and when, facilitating traceability and aiding in the investigation of security incidents or breaches. Moreover, establishing strong contractual agreements with vendors, encompassing comprehensive security provisions and compliance requirements, is critical for safeguarding data, intellectual property, and proprietary information.

By delineating security expectations, responsibilities, and liabilities in vendor contracts, companies can enforce stringent security measures across the supply chain and hold vendors accountable for adhering to agreed-upon standards. Lastly, fostering a culture of security awareness and vigilance among employees is essential for maintaining a strong defense against potential threats and vulnerabilities in the supply chain. By promoting a collaborative approach to security and encouraging employees to report any suspicious activity or deviations from security protocols, companies can enhance their overall resilience and mitigate the risk of supply chain-related security incidents.

Effective supply chain management requires a multifaceted approach that encompasses vendor assessment, secure delivery practices, documentation, contractual agreements, and employee awareness. By integrating these elements into their supply chain processes, web3 companies can enhance the security, resilience, and trustworthiness of their operations, ensuring the integrity of their products and services in an increasingly interconnected and dynamic business environment.

Chapter 21: Event Security

"Spies like conferences for the same reason everyone else does, free food, lots of information, and most people do not pay attention to another face."

Michael Weston, Burn Notice[451]

Corporate events such as conferences and meetups serve as invaluable platforms for networking, knowledge sharing, and business development. However, amidst the opportunities they present, these events also pose significant security risks, particularly in terms of corporate espionage and data breaches. Chapter 21 delves into the intricacies of event security within the web3 realm, emphasizing the critical need for proactive measures to safeguard sensitive information and mitigate potential threats.

Conferences play a central role in the web3 ecosystem, offering industry professionals a forum to exchange ideas, showcase innovations, and forge strategic partnerships. Yet, as these gatherings attract participants from diverse backgrounds and organizations, they become prime targets for espionage and information theft. Thus, it is imperative for companies to adopt stringent security protocols when participating in conferences, including the implementation of encryption measures for sensitive data transmission and adherence to strict access control policies to prevent unauthorized entry into restricted areas.

Similarly, meetup groups have emerged as instrumental platforms for fostering community engagement, facilitating collaboration, and promoting the adoption of web3 technologies. However, the open and informal nature of

[451] (2007). "Burn Notice." Paramount.

these gatherings can inadvertently expose companies to security vulnerabilities, such as unauthorized data disclosure or social engineering attacks. To mitigate these risks, organizations must prioritize employee training on data sensitivity, equipping staff with the knowledge and skills to discern and safeguard confidential information during interactions at meetups.

Ensuring physical security for data files and devices is paramount in mitigating the risk of theft or tampering at events. Companies should implement robust measures to secure laptops, smartphones, and other electronic devices used to access or store sensitive data, including password protection, encryption, and remote wiping capabilities. Additionally, the use of tamper-evident seals and secure storage containers can help prevent unauthorized access to physical documents and hardware.

It is a crucial aspect of event security is the vetting of attendees to verify their identities and affiliations, thereby reducing the likelihood of infiltration by malicious actors or competitors. Implementing stringent registration procedures, conducting background checks on participants, and employing facial recognition or biometric authentication technologies can enhance the integrity and trustworthiness of event attendance lists.

Event security in the web3 era demands a comprehensive and proactive approach to mitigate the myriad risks associated with corporate gatherings. By prioritizing data sensitivity training, implementing robust encryption measures, securing physical assets, and vetting attendees, companies can safeguard their proprietary information, preserve their reputation, and foster a secure environment conducive to collaboration and innovation in the dynamic landscape of web3.

Use of Conferences and Meetups in Web

Conferences and meetups stand as indispensable resources for web3 companies, offering unparalleled opportunities for knowledge acquisition, skill development, networking, and industry exposure. These events serve as vibrant hubs where professionals converge to share insights, showcase innovations, and cultivate meaningful connections within the web3 ecosystem. By participating in conferences and meetups, companies can gain valuable insights into emerging trends, technological advancements, and market dynamics, thereby enhancing their competitive edge and positioning themselves as thought leaders in the field.

One of the primary benefits of conferences and meetups lies in their capacity to provide access to a wealth of information and expertise across various domains within the web3 space. Through keynote speeches, panel discussions, workshops, and interactive sessions, attendees can gain invaluable insights from industry leaders, subject matter experts, and fellow practitioners.[452] These platforms offer a rich tapestry of perspectives, ideas, and best practices that can inspire innovation, inform strategic decision-making, and propel business growth in the rapidly evolving landscape of web3.

Conferences and meetups offer unparalleled networking opportunities, enabling professionals to connect with like-minded individuals, potential collaborators, investors, and clients. By engaging in meaningful conversations, exchanging business cards, and participating in networking sessions, attendees can expand their professional network,

---

[452] Bambacht, J. and Pouwelse, J., 2022. Web3: A decentralized societal infrastructure for identity, trust, money, and data. *arXiv preprint arXiv:2203.00398*.

forge strategic partnerships, and cultivate relationships that may lead to future collaborations or business ventures. Networking at these events fosters a sense of community and camaraderie within the web3 ecosystem, creating a supportive environment where ideas flourish and collaborations thrive.

To maximize the usefulness and exposure at conferences and meetups, web3 companies can adopt several best practices tailored to their specific objectives and target audience. Firstly, thorough pre-event preparation is essential, including researching the agenda, identifying key speakers and sessions of interest, and setting clear objectives for participation. By having a strategic plan in place, companies can optimize their time and resources, ensuring that they attend sessions relevant to their business goals and areas of focus.

Additionally, active participation in panel discussions, Q&A sessions, and networking events can enhance visibility and credibility within the industry. Companies should encourage their representatives to engage thoughtfully with fellow attendees, share their insights and experiences, and actively contribute to discussions. By positioning themselves as active contributors to the conversation, companies can establish themselves as trusted voices within the web3 community and attract attention to their brand and offerings.

Leveraging social media platforms and digital channels can amplify the reach and impact of participation in conferences and meetups. Companies can live-tweet key insights, share highlights from sessions, and engage with attendees and followers online, thereby extending the reach of their message beyond the confines of the event venue. By harnessing the power of digital communication, companies can amplify their presence, engage with a broader audience, and foster meaningful connections in the virtual sphere.

Realistically, companies should prioritize follow-up and relationship-building activities post-event to capitalize on the connections made and insights gained during the conference or meetup. Sending personalized follow-up emails, connecting with new contacts on professional networking platforms, and scheduling one-on-one meetings or calls can help solidify relationships and lay the groundwork for future collaborations or partnerships. By nurturing relationships over time, companies can cultivate a robust network of supporters, advocates, and collaborators within the web3 community, driving sustained growth and success in the dynamic landscape of decentralized technologies.

Conferences and meetups serve as invaluable platforms for web3 companies to access knowledge, expand their network, and showcase their expertise and offerings. By adopting best practices such as thorough preparation, active participation, leveraging digital channels, and prioritizing relationship-building activities, companies can maximize the usefulness and exposure at these events, establishing themselves as leaders in the web3 space and driving continued innovation and growth in the industry.

Data Training for Employees at Events (What to share and what not to share)

Training employees on what to say and what not to say at conferences is paramount to safeguarding the security of the company and protecting sensitive information from inadvertent disclosure or exploitation by malicious actors. Given the dynamic and collaborative nature of conferences, employees must exercise discretion and adhere to established protocols to mitigate potential risks and uphold the company's security posture. Here are ten key reasons why such training is vital, along with best practices for effective implementation:

- **Protection of Sensitive Information**: Employees need to understand the significance of safeguarding sensitive information, including proprietary data, intellectual property, and customer details, to prevent unauthorized access or exploitation.[453] Training should emphasize the importance of confidentiality and the potential consequences of disclosing sensitive information to unauthorized individuals.
- **Mitigation of Insider Threats**: Educating employees about the risks of insider threats and the tactics employed by malicious actors to exploit vulnerabilities can help mitigate the risk of internal data breaches or corporate espionage.[454] Training should raise awareness about common social engineering techniques used to manipulate individuals into divulging sensitive information.
- **Compliance with Regulatory Requirements**: Many industries are subject to strict regulatory requirements governing data privacy and security.[455]

---

[453] Stach, C., Gritti, C., Bräcker, J., Behringer, M. and Mitschang, B., 2022. Protecting sensitive data in the information age: State of the art and future prospects. *Future Internet*, *14*(11), p.302.
[454] Liu, L., De Vel, O., Han, Q.L., Zhang, J. and Xiang, Y., 2018. Detecting and preventing cyber insider threats: A survey. *IEEE Communications Surveys & Tutorials*, *20*(2), pp.1397-1417.
[455] Churchill, R.R. and Ulfstein, G., 2000. Autonomous institutional arrangements in multilateral environmental agreements: A little-noticed phenomenon in international law. *American Journal of International Law*, *94*(4), pp.623-659.

Employees must be educated about relevant regulations such as GDPR, HIPAA, or CCPA and instructed on how to handle sensitive information in compliance with legal obligations. Training should highlight the implications of non-compliance and the potential legal ramifications for the company.

- **Preservation of Corporate Reputation**: Inappropriate or careless remarks made by employees at conferences can have far-reaching implications for the company's reputation and brand image.[456] Training should emphasize the importance of maintaining professionalism and integrity in all interactions, both online and offline, to safeguard the company's reputation and build trust with stakeholders.
- **Protection Against Social Engineering Attacks**: Conferences provide fertile ground for social engineering attacks, where malicious actors may attempt to gather intelligence or exploit unsuspecting attendees for nefarious purposes.[457] Employees should be trained to recognize and respond to social engineering tactics, such as pretexting, phishing, or elicitation techniques, to thwart potential attacks.
- **Adherence to Corporate Policies**: Employees need clear guidance on corporate policies regarding information sharing, media interactions, and engagement with external parties. Training should reinforce adherence to company policies and procedures, including guidelines for media inquiries, public statements, and social media

---

[456] Rindova, V.P., 1997. Part VII: Managing reputation: Pursuing everyday excellence: The image cascade and the formation of corporate reputations. *Corporate reputation review*, *1*, pp.188-194.
[457] Salahdine, F. and Kaabouch, N., 2019. Social engineering attacks: A survey. *Future internet*, *11*(4), p.89.

engagement, to ensure consistency and compliance across the organization.[458]

- **Identification of Security Risks**: Effective training empowers employees to identify and report security risks or suspicious behavior observed during conferences.[459] By encouraging a culture of vigilance and accountability, employees can play an active role in detecting and addressing potential security threats before they escalate into serious incidents.
- **Promotion of Cyber Hygiene**: Cybersecurity awareness training should encompass fundamental principles of cyber hygiene, including password security, device encryption, and safe browsing practices.[460] Employees should be educated on the importance of strong passwords, the risks of using unsecured networks, and the need for regular software updates to mitigate vulnerabilities.
- **Role-specific Training**: Tailoring training programs to address the specific roles and responsibilities of employees ensures relevance and effectiveness.[461] Sales representatives, for example, may require training on handling customer inquiries and protecting sensitive sales data, while technical staff may need advanced training on secure coding practices and system configuration.

---

[458] *Id.*
[459] Pan, L. and Tomlinson, A., 2016. A systematic review of information security risk assessment. *International Journal of Safety and Security Engineering*, *6*(2), pp.270-281.
[460] Ncubukezi, T., Mwansa, L. and Rocaries, F., 2020, December. A review of the current cyber hygiene in small and medium-sized businesses. In *2020 15th International Conference for Internet Technology and Secured Transactions (ICITST)* (pp. 1-6). IEEE.
[461] *Id.*

- **Continual Education and Reinforcement**: Security awareness training should be an ongoing process rather than a one-time event, with regular updates and reinforcement activities to keep employees informed about evolving threats and best practices.[462] Interactive workshops, simulated phishing exercises, and real-world case studies can enhance engagement and retention of key concepts.

Training employees on what to say and what not to say at conferences is a critical component of an organization's cybersecurity strategy, providing frontline defense against insider threats, social engineering attacks, and reputational risks. By instilling a culture of security awareness, promoting compliance with corporate policies, and equipping employees with the knowledge and skills to identify and mitigate security risks, companies can enhance their resilience to emerging threats and protect their valuable assets in an increasingly complex and interconnected digital landscape.

Physical Security

The presence of physical security at conferences and meetups serves as a crucial layer of defense against potential exploitation or security breaches, complementing the efforts of employees to uphold data security protocols. Here are key reasons highlighting the importance of physical security measures:

- **Prevention of Unauthorized Access**: Physical security personnel play a vital role in controlling access to conference venues and restricted areas, ensuring that only authorized individuals are

[462] *Id.*

allowed entry. By implementing access control measures such as badge checks, identity verification, and perimeter security, physical security personnel help mitigate the risk of unauthorized individuals gaining access to sensitive information or high-security areas.

- **Deterrence of Criminal Activity**: The visible presence of uniformed security personnel acts as a deterrent to criminal activity, dissuading potential perpetrators from attempting theft, espionage, or other illicit activities. Their vigilant presence serves as a proactive deterrent, discouraging would-be attackers or opportunistic individuals from exploiting vulnerabilities or targeting unsuspecting attendees.
- **Response to Security Incidents**: In the event of security incidents or emergencies, physical security personnel are trained to respond swiftly and effectively, minimizing the impact and restoring order. Whether addressing disruptive behavior, handling suspicious packages, or coordinating with law enforcement agencies, their proactive approach enhances the safety and security of attendees and assets.
- **Protection of Valuable Assets**: Conferences and meetups often involve the presence of valuable assets such as electronic devices, promotional materials, and proprietary information. Physical security measures such as surveillance cameras, secure storage areas, and asset tracking systems help safeguard these assets against theft, loss, or unauthorized access, preserving their integrity and value to the organization.

The deployment of physical security at conferences and meetups is essential to reinforce the overall security posture of web3 companies and mitigate the risk of exploitation or compromise. By maintaining a visible presence, deterring criminal activity, responding to incidents, and protecting valuable assets, physical security personnel contribute to creating a safe and secure environment conducive to productive networking, collaboration, and knowledge sharing.

Attendee Vetting (if possible)

Attendee vetting, when ethically and legally feasible, offers significant advantages to web3 companies in maintaining the security and integrity of their events. Thorough vetting processes allow organizations to assess the backgrounds and identities of attendees, reducing the risk of unauthorized access, disruptive behavior, or malicious intent. By implementing stringent attendee vetting measures, companies can create a safer and more secure environment for participants, speakers, and organizers, enhancing the overall experience and reputation of the event.

Vetting attendees helps mitigate potential security threats by identifying individuals with a history of misconduct, criminal activity, or disruptive behavior. Through background checks and verification procedures, companies can screen out individuals who may pose a risk to the event's security or undermine its objectives.[463] By proactively identifying and addressing security concerns, organizations can minimize the likelihood of incidents such

---

[463] Velin, L., Lartigue, J.W., Johnson, S.A., Zorigtbaatar, A., Kanmounye, U.S., Truche, P. and Joseph, M.N., 2021. Conference equity in global health: a systematic review of factors impacting LMIC representation at global health conferences. *BMJ Global Health*, *6*(1), p.e003455.

as fraud, theft, or harassment, safeguarding the well-being and interests of all participants.

In addition to enhancing security, attendee vetting can also facilitate networking and collaboration by ensuring that participants are aligned with the event's goals, values, and target audience. By curating a diverse and qualified attendee base, companies can create opportunities for meaningful engagement, knowledge sharing, and professional development. Vetting processes can be tailored to match the specific requirements and objectives of each event, allowing organizers to customize the attendee experience and maximize the value proposition for all participants.

Attendee vetting plays a crucial role in compliance with regulatory requirements and industry standards governing event security and risk management. By conducting due diligence on attendees and verifying their identities, organizations can demonstrate a commitment to upholding legal and ethical standards, mitigating potential liabilities and reputational risks. Compliance with data protection regulations and privacy laws is paramount when collecting and processing attendee information, and vetting procedures should be designed and implemented in accordance with relevant legal frameworks.

Attendee vetting serves as a proactive risk mitigation strategy, enabling companies to anticipate and address security threats before they escalate into incidents. By investing in robust vetting processes and security protocols, web3 companies can foster a culture of safety, trust, and professionalism at their events, enhancing the overall experience for attendees and contributing to the success and reputation of the organization.

Conclusion

In conclusion, ensuring the security of events in the web3 landscape demands a comprehensive approach that addresses various potential threats and vulnerabilities. From conferences to meetups, companies must prioritize security measures to protect sensitive information, foster a safe environment, and uphold their reputation. By leveraging attendee vetting, organizations can proactively mitigate risks, screen out potential threats, and enhance the overall security posture of their events.

Furthermore, effective employee training on appropriate behavior and communication at events is essential to minimize security risks and maintain professionalism. By educating staff on the importance of discretion, data sensitivity, and compliance with security protocols, companies can empower their employees to act as ambassadors of security and uphold the organization's standards during events.

Physical security measures, such as access controls and surveillance, play a crucial role in safeguarding event venues and preventing unauthorized access or disruptions. By implementing robust physical security protocols, companies can create a secure environment for attendees, speakers, and staff, reducing the risk of incidents and ensuring the smooth operation of the event.

Overall, by adopting a holistic approach to event security that encompasses attendee vetting, employee training, and physical security measures, web3 companies can mitigate risks, protect sensitive information, and uphold the integrity of their events. Through proactive risk management and adherence to best practices, organizations can create safe, inclusive, and successful events that contribute to their overall goals and objectives in the digital landscape.

Chapter 22: Insurance Coverage

"There are worse things in life than death, have you ever spent an evening with an insurance salesman."

Woody Allen[464]

Chapter 22 examins the critical importance of insurance coverage for web3 companies operating in today's dynamic and evolving digital landscape. In an era marked by unprecedented technological advancements and complex cybersecurity threats, having robust insurance policies in place is paramount to safeguarding against potential risks and mitigating financial liabilities. This chapter explores various types of insurance coverage essential for web3 companies, including corporate insurance, errors and omissions (E&O) insurance, vendor insurance, contractor insurance, and travel insurance for employees.

Corporate insurance serves as a foundational component of risk management for web3 companies, providing financial protection against a wide range of potential liabilities, including property damage, bodily injury, and legal expenses. By securing comprehensive corporate insurance coverage, organizations can mitigate the financial impact of unforeseen events and safeguard their assets, operations, and reputation in an increasingly litigious business environment.

Errors and omissions (E&O) insurance is specifically designed to protect web3 companies from liability arising from professional services or advice provided to clients. As the digital landscape continues to evolve and companies engage in complex transactions and projects, the risk of

---

[464] (2024). "Insurance Quotes, Woody Allen." Brany Quote at https://www.brainyquote.com/topics/insurance-quotes.

errors, omissions, or professional negligence increases. E&O insurance provides vital protection against legal claims alleging financial loss or damage due to errors in professional services, software implementation, or other business activities.

Insurance coverage for vendors and contractors is another essential consideration for web3 companies, especially those that rely on external partners or suppliers to support their operations. By ensuring that vendors and contractors maintain adequate insurance coverage, companies can mitigate the risk of potential liabilities stemming from third-party actions, errors, or accidents. Additionally, requiring vendors and contractors to carry insurance coverage helps protect the company's interests and ensures compliance with contractual obligations.

Travel insurance for employees is crucial for web3 companies with a global presence or employees who frequently travel for business purposes. Travel insurance provides coverage for medical emergencies, trip cancellations, lost luggage, and other unforeseen events that may occur during business travel. By offering comprehensive travel insurance benefits, companies can prioritize the safety and well-being of their employees while mitigating financial risks associated with travel-related incidents.

Corporate Insurance

Corporate insurance, particularly in the form of an umbrella policy, plays a pivotal role in bolstering security measures for web3 companies. Firstly, an umbrella policy provides an additional layer of protection beyond the limits of primary liability insurance policies, such as general liability or professional liability coverage. In the context of security, this extended coverage can prove invaluable in mitigating the financial fallout from unforeseen security breaches, cyberattacks, or data breaches that may result in costly legal claims, damages, or regulatory fines.

An umbrella policy offers comprehensive coverage across multiple liability areas, including property damage, bodily injury, legal expenses, and reputational harm. In the realm of security, where companies face diverse and evolving risks ranging from physical breaches to cyber threats, having broad coverage under an umbrella policy can help address a wide spectrum of potential liabilities.[465] This comprehensive approach ensures that web3 companies are adequately protected against various security-related risks and can respond effectively to unexpected security incidents.

Additionally, corporate insurance serves as a critical risk management tool by providing financial protection against the fallout from security incidents or data breaches. In the event of a security breach, web3 companies may incur substantial expenses related to forensic investigations, legal defense costs, notification of affected parties, and potential regulatory fines. An umbrella policy can help cover these expenses, thereby mitigating the financial impact on the company and facilitating a more efficient response to security incidents.

Umbrella policies can offer peace of mind to web3 companies by providing coverage for potential gaps or exclusions in other insurance policies. As security threats continue to evolve and become more sophisticated, traditional insurance policies may not fully address all

---

[465] Petrescu, R.M., 2016. Identifying best practices in insurance. *Challenges of the knowledge society*, p.232.

aspects of cyber risk or emerging security challenges. An umbrella policy can fill these gaps and provide additional protection against a wide range of security-related liabilities, ensuring that web3 companies have comprehensive insurance coverage tailored to their unique security needs.

Corporate insurance, particularly in the form of an umbrella policy, serves as a critical safeguard for web3 companies navigating the complex landscape of security risks and challenges. By providing broad coverage, financial protection, and peace of mind, an umbrella policy enables companies to enhance their security posture, mitigate potential liabilities, and safeguard their assets, operations, and reputation in an increasingly volatile and interconnected digital environment.

Errors and Omissions

Errors and omissions (E&O) insurance is a crucial component of a comprehensive risk management strategy for web3 companies, particularly those operating in industries where professional services or advice are integral to their business activities. This type of insurance provides protection against claims arising from errors, negligence, or failures to perform professional duties, which could result in financial losses for clients or third parties.[466] The importance of E&O insurance lies in its ability to shield web3 companies from the potentially devastating financial consequences of litigation or legal claims alleging professional negligence or mistakes.

---

[466] Ellison, J.N. and Frankel, R.E., 1998. A guide to E&O insurance. *Risk Management*, *45*(6), p.33.

One key aspect of E&O insurance is its role in safeguarding web3 companies against the costs associated with legal defense and settlements in the event of a lawsuit alleging errors or omissions in the delivery of professional services. Legal disputes arising from alleged mistakes, oversights, or failures to meet client expectations can be protracted and expensive, consuming valuable resources and undermining the financial stability of the company. E&O insurance provides financial support to cover legal expenses, including attorney fees, court costs, and settlement payments, thereby helping web3 companies navigate legal challenges and protect their assets.

E&O insurance enhances the credibility and reputation of web3 companies by demonstrating their commitment to professionalism, quality assurance, and risk management. By carrying E&O coverage, companies signal to clients, partners, and stakeholders that they take their responsibilities seriously and are prepared to address any potential errors or omissions that may occur in the course of their business activities.[467] This can instill confidence in clients and mitigate concerns about potential liabilities, thereby strengthening relationships and fostering trust in the company's ability to deliver reliable and high-quality services.

Another key benefit of E&O insurance is its role in protecting the financial interests of web3 companies by providing a safety net against unexpected liabilities and financial losses. In the fast-paced and complex world of web3 technologies, where innovation and disruption are constant, the risk of unintentional errors, misinterpretations, or oversights is inherent. E&O insurance helps mitigate these risks by offering financial protection against claims of

[467] *Id.*

professional negligence or mistakes, allowing companies to focus on their core business activities without fear of catastrophic financial repercussions.

Overall, E&O insurance is an essential risk management tool for web3 companies seeking to protect themselves against potential liabilities stemming from professional errors or omissions. By providing financial support for legal defense, settlements, and damages, E&O insurance helps mitigate the financial risks associated with legal disputes and lawsuits, enhances the company's credibility and reputation, and safeguards its financial interests in an increasingly competitive and litigious business environment.

Vendors and Contractor Insurance

Ensuring that vendors/Contractors carry adequate insurance coverage is a prudent risk management practice for web3 companies seeking to protect their interests and mitigate potential liabilities associated with outsourced services or products. By requiring vendors and contractors to maintain appropriate insurance policies and providing proof of coverage, companies can safeguard themselves against financial losses arising from vendor-related incidents or accidents, thereby enhancing their overall risk management strategy and protecting their bottom line.[468]

One compelling reason to verify a vendor's or contractor's insurance coverage is to shift the financial burden of liability away from the company and onto the vendor in the event of accidents, damages, or other unforeseen events. By stipulating insurance requirements in vendor contracts and mandating proof of coverage, companies can transfer the

[468] Sherman, *supra.*

responsibility for potential liabilities to the vendor or contractor, reducing their own exposure to financial risks and ensuring that any damages or losses incurred are covered by the vendor's insurance policy.

Additionally, requiring vendors and contractors to carry insurance demonstrates a commitment to due diligence and risk management, signaling to stakeholders, clients, and regulatory authorities that the company takes its obligations seriously and prioritizes the protection of its interests. This proactive approach to 3rd party risk management can enhance the company's reputation, build trust with clients and partners, and mitigate concerns about potential liabilities or disruptions associated with vendor relationships.

Verifying vendor/contractor insurance coverage helps ensure compliance with legal and regulatory requirements governing liability and insurance in various industries. Many jurisdictions impose minimum insurance requirements for vendors and contractors engaged in certain activities or services, and failure to comply with these requirements can result in legal penalties, fines, or regulatory sanctions.[469] By verifying vendor/contractor insurance coverage and ensuring compliance with applicable laws and regulations, companies can avoid potential legal pitfalls and protect themselves from costly enforcement actions.

Further, requiring vendors and contractors to submit proof of insurance can help clarify the scope of coverage and identify any gaps or deficiencies in insurance policies that may pose risks to the company. By reviewing the terms and conditions of vendor insurance policies, companies can

---

[469] *Id.*

assess the adequacy of coverage, verify that it aligns with contractual obligations and project requirements, and address any discrepancies or concerns before entering into agreements with vendors. This proactive risk assessment enables companies to make informed decisions about vendor selection, negotiate appropriate insurance provisions, and mitigate potential risks associated with vendor relationships.

Verifying vendor and contractor insurance coverage and requiring vendors/contractors to submit proof of insurance are essential risk management practices for web3 companies seeking to protect their interests, minimize liabilities, and ensure compliance with legal and regulatory requirements. By shifting the financial burden of liability to vendors, demonstrating due diligence and compliance with industry standards, and proactively assessing and addressing insurance-related risks, companies can enhance their overall risk management strategy and safeguard their operations in an increasingly complex and interconnected business environment.

Travel Insurance

Travel insurance plays a crucial role in safeguarding the interests of web3 companies that frequently send their employees to events, conferences, or business trips. These policies provide financial protection against a wide range of risks and uncertainties associated with travel, offering peace of mind to both employees and employers alike.[470] One of the primary reasons why travel insurance is essential for web3 companies is its ability to mitigate the financial impact of unexpected events or emergencies that

[470] Leggat, P.A., Carne, J. and Kedjarune, U., 1999. Travel insurance and health. *J Travel Med*, *6*(4), pp.243-248.

may occur during travel, such as trip cancellations, flight delays, or medical emergencies.

For web3 companies, whose operations often involve extensive travel to attend conferences, meet with clients, or participate in industry events, travel insurance serves as a safety net against potential disruptions or unforeseen circumstances that could disrupt business activities.[471] By providing coverage for trip cancellations or interruptions due to reasons beyond the company's control, such as illness, injury, or natural disasters, travel insurance helps minimize financial losses and allows employees to focus on their professional duties without worrying about the financial repercussions of travel-related incidents.

Sometimes travel insurance offers valuable protection against medical emergencies that may arise while employees are traveling for business purposes. In the event of illness or injury requiring medical attention during a trip, travel insurance can cover medical expenses, hospitalization costs, and emergency medical evacuation, ensuring that employees receive prompt and adequate medical care without facing exorbitant out-of-pocket expenses.[472] This aspect of travel insurance is particularly important for web3 companies whose employees frequently travel to destinations where access to quality healthcare services may be limited or costly.

Additionally, travel insurance provides coverage for lost or stolen belongings, including personal belongings, electronic devices, and business equipment, offering financial compensation for the value of the items lost or

[471] Leggat, P.A., Carne, J. and Kedjarune, U., 1999. Travel insurance and health. *J Travel Med*, *6*(4), pp.243-248.
[472] *Id.*

damaged during travel. This aspect of travel insurance helps mitigate the risks associated with theft, loss, or damage to valuable assets, ensuring that employees can conduct their business activities without the fear of financial loss due to unforeseen incidents.

Furthermore, travel insurance often includes benefits such as emergency assistance services, which provide round-the-clock support and guidance to travelers in distress. Whether it's arranging emergency medical treatment, coordinating travel arrangements, or providing legal or logistical assistance, these services can be invaluable for web3 companies whose employees may encounter challenges or emergencies while traveling in unfamiliar environments.

In conclusion, travel insurance is a vital risk management tool for web3 companies that frequently send their employees on business trips or travel to events. By providing financial protection against trip cancellations, medical emergencies, lost belongings, and other travel-related risks, travel insurance helps mitigate the financial impact of unforeseen events and ensures the safety and well-being of employees while they are away on business. Incorporating travel insurance into their risk management strategy allows web3 companies to protect their interests, support their employees, and maintain business continuity in an increasingly mobile and globalized world.

Conclusion

Insurance coverage plays a pivotal role in bolstering the security posture of web3 companies, offering financial protection against a myriad of risks and liabilities inherent in their operations. Through comprehensive insurance policies tailored to the specific needs of web3 enterprises, companies can safeguard their interests, assets, and personnel against unforeseen events, ensuring business continuity and resilience in the face of challenges. From corporate insurance to errors and omissions coverage, insurance serves as a critical risk management tool, providing a safety net for web3 companies navigating the complex and dynamic landscape of the digital economy.

Corporate insurance, including umbrella policies, offers broad-ranging protection against various liabilities, lawsuits, and damages that may arise in the course of business operations. By consolidating multiple insurance coverages under a single policy, web3 companies can streamline their risk management efforts and minimize gaps in coverage, enhancing their overall security posture and financial stability. Moreover, errors and omissions insurance provides essential protection against claims of professional negligence or inadequate performance, shielding web3 firms from costly litigation and reputational damage.

In addition to protecting the company itself, insurance coverage extends to vendors, contractors, and employees, ensuring that all parties involved in business activities are adequately covered against potential risks and liabilities. By requiring vendors and contractors to carry insurance and verifying the validity of their policies, web3 companies can mitigate the risk of liability transfer and financial exposure, fostering a more secure and resilient supply chain ecosystem. Similarly, travel insurance for employees traveling on behalf of the company offers vital protection against travel-related risks and emergencies, safeguarding the well-being of personnel and minimizing disruptions to business operations.

Insurance coverage serves as a strategic investment in risk mitigation and business continuity, enabling web3 companies to navigate uncertainties and challenges with confidence. By proactively addressing potential risks and vulnerabilities through comprehensive insurance solutions, organizations can mitigate financial losses, protect their assets, and uphold their reputation in the marketplace. Incorporating insurance into the company's risk management strategy demonstrates a commitment to safeguarding stakeholders' interests and ensuring the long-term sustainability of the business in a rapidly evolving digital landscape.

In essence, insurance coverage is an indispensable component of the security infrastructure for web3 companies, providing essential financial protection, risk mitigation, and peace of mind in an increasingly complex and interconnected business environment. By leveraging insurance solutions tailored to their specific needs and risk profiles, web3 enterprises can fortify their resilience, enhance their competitiveness, and thrive in the dynamic and evolving landscape of the digital economy.

Conclusion Part IV

As we conclude Part IV, it becomes evident that security in the web3 era is not merely an afterthought but a fundamental aspect of business operations in the digital age. Throughout this section, we have delved into various dimensions of cybersecurity and risk management, highlighting the critical importance of proactive measures, robust protocols, and comprehensive strategies to safeguard web3 companies against evolving threats and vulnerabilities. From hardware security to employee training, each chapter has underscored the multifaceted nature of cybersecurity, emphasizing the need for a holistic approach that encompasses technology, policy, and human factors.

One of the overarching themes that emerge from Part IV is the interconnectedness of security across different facets of web3 enterprises. Whether it is securing hardware devices, protecting data centers, or managing third-party vendors, effective security measures require a coordinated effort across all levels of the organization. By integrating security into every aspect of business operations, from procurement to employee training, web3 companies can create a culture of security that permeates throughout the organization, fostering resilience and adaptability in the face of emerging threats.

Furthermore, the chapters in Part IV have highlighted the importance of staying ahead of the curve in a rapidly evolving threat landscape. As cyber threats continue to evolve in sophistication and complexity, web3 companies must remain vigilant and proactive in their approach to security. This entails not only investing in state-of-the-art technologies and security solutions but also fostering a culture of continuous learning and adaptation, where employees are empowered to identify and respond to security threats effectively.

The role of compliance and regulation cannot be understated in the context of web3 security. With an increasing number of regulations and standards governing data privacy and cybersecurity, web3 companies must ensure compliance with relevant laws and regulations to mitigate legal and reputational risks. By aligning their security practices with industry best practices and regulatory requirements, organizations can demonstrate their commitment to protecting customer data and upholding trust in the digital ecosystem.

Another key takeaway from Part IV is the importance of collaboration and partnerships in enhancing security resilience. In an interconnected digital ecosystem, no single organization can address all security challenges alone. By fostering partnerships with industry peers, sharing threat intelligence, and collaborating with cybersecurity experts, web3 companies can leverage collective expertise and resources to strengthen their security posture and mitigate emerging threats effectively.

Part IV has provided valuable insights and actionable strategies for web3 companies seeking to enhance their security resilience in an increasingly complex and dynamic digital landscape. By adopting a proactive and comprehensive approach to cybersecurity, leveraging state-of-the-art technologies, and fostering a culture of security throughout the organization, web3 enterprises can mitigate risks, protect assets, and thrive in the digital economy. As the digital ecosystem continues to evolve, embracing security as a core business imperative will be essential for web3 companies to succeed and thrive in the ever-changing landscape of the digital era.

Part V: Digital Identity and Ownership

Introduction to Part V

Part V of this comprehensive guide engages security in the web3 landscape, where digital innovations intersect with evolving cybersecurity threats. As businesses increasingly leverage decentralized technologies and embrace the principles of blockchain, they are confronted with a myriad of security challenges that demand innovative solutions and proactive measures to safeguard sensitive information and mitigate risks effectively.

In this section, we explore a diverse range of topics spanning from the intricacies of anonymity and digital identity to the complexities of authentication protocols and access controls. Each chapter offers valuable insights into the dynamic landscape of web3 security, providing practical guidance and strategic recommendations to help organizations navigate the ever-changing threat landscape and fortify their cyber defenses.

Chapter 23 looks into the enigmatic world of anon culture in web3, shedding light on the challenges of implementing Know Your Customer (KYC) protocols in anonymous environments. By examining the nuances of anonymity and accountability, this chapter explores the delicate balance between privacy and transparency, highlighting the importance of establishing robust security measures to maintain trust and integrity in decentralized ecosystems.

In Chapter 24, we explore the critical role of KYC protocols in combating financial crimes such as money laundering, terrorism financing, and human trafficking. By delving into the regulatory landscape and discussing best practices for KYC compliance, this chapter offers practical guidance to organizations seeking to mitigate legal and reputational risks while contributing to global efforts to combat illicit activities.

Chapter 25 shifts the focus to the pervasive threats of Sybil attacks and identity fraud, emphasizing the importance of proactive measures to protect digital identities and prevent fraudulent activities. Through a comprehensive analysis of authentication mechanisms and fraud detection technologies, this chapter equips organizations with the knowledge and tools needed to enhance security and mitigate risks effectively.

In Chapter 26, we examine the intersection of accessibility and security in digital identity management, exploring strategies to empower users while maintaining strong security measures. By promoting a culture of security awareness and implementing user-centric authentication solutions, organizations can strike a balance between usability and protection, fostering trust and confidence in their digital ecosystems.

Throughout this section, we emphasize the importance of collaboration, innovation, and continuous improvement in addressing the evolving challenges of web3 security. By leveraging cutting-edge technologies, adopting best practices, and fostering partnerships with industry stakeholders, organizations can enhance their resilience and adaptability in the face of emerging threats, ensuring the integrity, confidentiality, and availability of their digital assets.

Chapter 23: Web3 and the Anons

"The internet made fame wack and anonymity cool."

Frank Ocean[473]

In Web3, anonymity holds both promise and peril. This chapter looks into the interplay between the culture of anonymity, Know Your Customer (KYC) regulations, and the ethical implications for businesses operating in this decentralized digital realm. As companies navigate the complex terrain of Web3, understanding the dynamics of anonymity becomes paramount to maintaining security and integrity.

Anonymity has long been a hallmark of internet culture, offering users freedom of expression and privacy. However, in the context of Web3, where decentralized platforms and cryptocurrencies reign supreme, anonymity takes on new dimensions. It becomes not just a matter of personal preference but a fundamental aspect of digital identity and transactions. This chapter explores how anonymity shapes interactions within Web3 communities and the challenges it poses for businesses seeking to establish trust and accountability.

KYC regulations, designed to combat money laundering and terrorist financing, present a formidable hurdle for Web3 companies embracing anonymity. The tension between regulatory compliance and user privacy underscores the need for innovative solutions that strike a delicate balance between security and individual rights. This chapter examines the evolving landscape of KYC requirements in the Web3 space and the strategies

[473] (2024). "Anonymity Quotes, Frank Ocean." Brainy Quote at https://www.brainyquote.com/topics/anonymity-quotes.

companies employ to reconcile regulatory obligations with user expectations.

One prevailing approach adopted by Web3 companies is the "we know, others don't" policy, which encapsulates the concept of selective disclosure of user identities. By safeguarding user data while ensuring compliance with legal mandates, companies aim to foster trust and transparency within their ecosystems. This policy raises crucial ethical questions regarding the extent of corporate responsibility and the rights of individuals to remain anonymous online.

As Web3 continues to redefine the boundaries of digital interactions, businesses must grapple with the implications of anonymity for security, governance, and user empowerment. This chapter invites readers to explore the nuanced dynamics of Web3 anonymity and its implications for the future of decentralized systems and online communities. By gaining a deeper understanding of these issues, companies can navigate the complexities of Web3 with confidence and integrity.

The Anon Culture in Web3

In the decentralized landscape of Web3, anonymity is not merely a feature but a cultural cornerstone. Anon culture, rooted in the ethos of privacy and individual sovereignty, permeates online interactions and transactions. While anonymity empowers users to engage freely and pseudonymously, it also presents significant challenges for security and accountability.[474]

---

[474] Tunstall, K.E., 2011. " You're Either Anonymous or You're Not!": Variations on Anonymity in Modern and Early Modern Culture. *MLN*, *126*(4), pp.671-688.

One of the primary challenges posed by anon culture in Web3 is the difficulty of establishing trust and verifying identities. Unlike traditional centralized systems where users are often required to undergo identity verification, Web3 platforms operate on principles of pseudonymity and cryptographic anonymity.[475] While this fosters a sense of privacy and autonomy, it also creates opportunities for malicious actors to exploit the veil of anonymity for illicit activities such as fraud, hacking, and identity theft.

The decentralized nature of Web3 introduces complexities in security protocols and governance mechanisms. Without a central authority to enforce compliance or monitor user behavior, ensuring the integrity and security of transactions becomes inherently challenging.[476] Anon culture amplifies these challenges by blurring the lines between legitimate users and malicious actors, making it difficult to discern friend from foe in cyberspace.

Another aspect of anon culture that complicates security in Web3 is the prevalence of pseudonymous accounts and the lack of real-world identities tied to digital personas. While pseudonymity protects user privacy, it also hampers efforts to hold individuals accountable for their actions. In cases of cyberattacks or fraudulent activities, tracing perpetrators and enforcing legal recourse becomes arduous, if not impossible, in the absence of identifiable information.

Overall, anon culture in Web3 presents a multifaceted security challenge that demands innovative solutions and

---

[475] Auerbach, D., 2012. Anonymity as culture: Treatise. *Triple Canopy*, *15*(09.02).

[476] McDonald, K., 2015. From Indymedia to Anonymous: rethinking action and identity in digital cultures. *Information, Communication & Society*, *18*(8), pp.968-982.

robust security measures. Balancing the principles of privacy and anonymity with the need for accountability and trust is paramount in safeguarding the integrity and resilience of decentralized systems.[477] As Web3 continues to evolve, addressing the complexities of anon culture will be crucial in building a secure and trustworthy digital ecosystem for all stakeholders involved.

KYC Issues

Know Your Customer (KYC) regulations are designed to mitigate risks such as money laundering, fraud, and terrorist financing by verifying the identities of individuals engaging in financial transactions.[478] However, in the realm of Web3, where anon culture prevails, implementing KYC measures becomes inherently challenging due to several factors.

First, anon culture promotes pseudonymity and privacy, allowing users to transact without revealing their true identities.[479] This anonymity undermines the effectiveness of traditional KYC procedures, which rely on verifying the real-world identities of individuals. Without access to identifiable information, businesses and platforms struggle to comply with regulatory requirements and perform thorough KYC checks.

Second, the decentralized nature of Web3 platforms complicates KYC processes by eliminating centralized

---

[477] Auerbach, *supra*.
[478] Parra Moyano, J. and Ross, O., 2017. KYC optimization using distributed ledger technology. *Business & Information Systems Engineering*, *59*, pp.411-423.
[479] Rajput, V.U., 2013. Research on know your customer (KYC). *International Journal of Scientific and Research Publications*, *3*(7), pp.541-546.

authorities or intermediaries responsible for identity verification.[480] In traditional financial systems, banks and financial institutions serve as gatekeepers for KYC compliance, but in Web3, there is no central entity to oversee or enforce such regulations. As a result, implementing KYC becomes decentralized and reliant on self-verification, raising concerns about the accuracy and reliability of user-provided information.

Anon culture fosters a distrust of centralized institutions and authorities, making users reluctant to disclose personal information for KYC purposes.[481] The ethos of privacy and individual sovereignty inherent in anon culture conflicts with the intrusive nature of KYC requirements, leading to resistance and non-compliance among users. This resistance further exacerbates the challenges of conducting effective KYC in Web3 environments.

Additionally, the global nature of Web3 transcends geographical boundaries, making it difficult to enforce uniform KYC standards across different jurisdictions. While some regions may have stringent KYC regulations, others may lack robust oversight or enforcement mechanisms, creating inconsistencies and loopholes that can be exploited by bad actors.[482] This lack of uniformity

---

480 Parate, S., Josyula, H.P. and Reddi, L.T., 2023. Digital identity verification: transforming KYC processes in banking through advanced technology and enhanced security measures. *International Research Journal of Modernization in Engineering Technology and Science*, *5*(9), pp.128-137.

481 *Id.*

482 Sinha, P. and Kaul, A., 2018. Decentralized KYC system. *International Research Journal of Engineering and Technology (IRJET)*, *5*(8), pp.1209-1210.

hampers KYC efforts and undermines the effectiveness of regulatory measures in combating financial crimes.

Overall, anon culture poses significant obstacles to KYC compliance in Web3, necessitating innovative approaches and technological solutions to address the inherent complexities.[483] Balancing the principles of privacy and anonymity with regulatory requirements for identity verification remains a formidable challenge in the evolving landscape of decentralized finance and digital interactions.

We Know, Others Don't Policy

The option of a company holding the name of an anonymous user without releasing it presents a delicate balance between privacy protection and accountability. In scenarios where anon culture prevails and users prioritize pseudonymity, platforms may adopt a policy of knowing the identity of users internally but refraining from disclosing it publicly. This approach allows companies to maintain compliance with regulatory requirements while respecting users' desires for anonymity.

Holding the name of an anonymous user introduces potential risks, particularly concerning the security of that information. In the event of a data breach or unauthorized disclosure, the user's anonymity could be compromised, leading to privacy violations and potential harm. To mitigate this risk, companies must implement robust security measures to safeguard sensitive user data, including encryption protocols, access controls, and regular security audits. Additionally, companies should establish clear policies and procedures for handling user information

[483] Mansoor, N., Antora, K.F., Deb, P., Arman, T.A., Manaf, A.A. and Zareei, M., 2023. A Review of Blockchain Approaches for KYC. *IEEE Access*.

responsibly and ethically, ensuring transparency and accountability in their practices.

In cases where an anonymous user's identity is doxed or unlawfully exposed, the company may explore options for providing recourse and protection to the affected individual. This may involve offering legal support or indemnification against potential consequences resulting from the unauthorized disclosure of their identity. By proactively supporting and advocating for the rights of anonymous users, companies can foster trust and confidence within their communities while upholding principles of privacy and security in the digital realm.

Conclusion

Navigating the complexities of anon culture within the Web3 landscape presents significant challenges for security and identity verification. The prevalence of pseudonymity and the desire for privacy among users contribute to a unique dynamic where traditional methods of KYC and identity verification may prove ineffective. As companies strive to balance user privacy with regulatory compliance and security requirements, they must develop innovative approaches to address these challenges.

The difficulty of verifying the identities of anonymous users underscores the importance of robust security measures and risk management strategies. Companies must implement stringent security protocols to protect user data and mitigate the risks associated with potential breaches or unauthorized disclosures. By investing in advanced encryption techniques, access controls, and comprehensive security frameworks, organizations can enhance their resilience against emerging threats in the Web3 ecosystem.

The concept of holding the name of an anonymous user while refraining from public disclosure introduces a nuanced ethical dilemma. While this approach may offer a compromise between privacy protection and accountability, it also necessitates careful consideration of the potential risks and implications involved. Companies must prioritize transparency, integrity, and user trust in their handling of sensitive information, ensuring that their practices align with ethical standards and legal requirements.

As Web3 continues to evolve and anon culture shapes the digital landscape, companies must adapt their security strategies and policies to effectively address emerging threats and safeguard user privacy. By embracing a proactive and collaborative approach to security, organizations can foster a culture of trust and accountability while navigating the complex interplay between anonymity, identity verification, and data protection in the digital age.

Chapter 24: KYC Protocols

"I don't know why people are so keen to put the details of their private life in public; they forget that invisibility is a superpower."

Banksy[484]

Chapter 24 breaks down the critical role of Know Your Customer (KYC) protocols in bolstering security within the Web3 landscape. KYC procedures serve as a cornerstone of regulatory compliance and risk mitigation, playing a pivotal role in safeguarding against illicit activities such as money laundering, terrorism financing, and human trafficking. As digital transactions become increasingly prevalent and complex, the need for robust KYC measures becomes more pronounced, underscoring the importance of thorough identity verification and due diligence processes.

One of the primary objectives of KYC protocols is to prevent and detect financial crimes, including money laundering and terrorist financing. By verifying the identities of customers and assessing the legitimacy of their transactions, financial institutions and businesses can mitigate the risk of inadvertently facilitating illicit activities. KYC procedures enable organizations to establish the identity of their customers, monitor their transactional behavior, and flag suspicious activities for further investigation, thereby enhancing the overall integrity of the financial system.

In addition to combating financial crimes, KYC protocols also play a crucial role in addressing broader societal

---

[484] (2024) "Anonymity Quotes, Banksy." Good Quote https://www.goodreads.com/quotes/tag/anonymity.

challenges, such as human trafficking and exploitation. By implementing rigorous identity verification measures, businesses can help prevent the illicit movement of funds associated with human trafficking networks and other forms of exploitation. Furthermore, KYC procedures empower organizations to identify and report suspicious activities that may be indicative of human trafficking or related offenses, thereby contributing to global efforts to combat these egregious crimes.

The adoption of effective KYC protocols is essential for maintaining regulatory compliance and upholding industry standards across various sectors. Regulatory authorities impose stringent requirements on financial institutions, cryptocurrency exchanges, and other entities to implement robust KYC measures as part of their anti-money laundering (AML) and counter-terrorism financing (CTF) efforts. By adhering to these regulatory mandates, organizations can mitigate legal risks, protect their reputations, and foster trust among stakeholders.

The emergence of innovative technologies, such as blockchain and digital identity solutions, presents both opportunities and challenges for KYC implementation. While blockchain technology offers enhanced security and transparency, it also introduces complexities related to identity management and data privacy. As organizations navigate the evolving regulatory landscape and leverage new technologies to enhance their KYC capabilities, they must strike a balance between innovation and compliance to effectively mitigate risks and ensure the integrity of their operations.

AML Considerations

Anti-Money Laundering (AML) considerations are paramount in KYC protocols, as they aim to prevent the illicit conversion of illegally obtained funds into legitimate assets. AML regulations require financial institutions and businesses to implement robust procedures for identifying and reporting suspicious transactions that may be indicative of money laundering activities.[485] By conducting thorough due diligence on customers and monitoring their transactional behavior, organizations can detect and deter attempts to launder money through their platforms.

One of the key aspects of AML considerations is the implementation of customer due diligence (CDD) measures, which involve verifying the identity of customers and assessing the nature of their transactions. Enhanced due diligence (EDD) may be required for high-risk customers or transactions, requiring additional scrutiny and documentation to ensure compliance with AML regulations.[486] By adhering to CDD and EDD requirements, organizations can mitigate the risk of inadvertently facilitating money laundering activities and protect themselves from regulatory scrutiny and penalties.

AML considerations encompass the development of robust transaction monitoring systems that enable organizations to detect suspicious activities in real-time. These systems analyze transactional data for anomalies and red flags, such as unusually large transactions, frequent cash deposits, or

---

[485] Kemal, M.U., 2014. Anti-money laundering regulations and its effectiveness. *Journal of Money Laundering Control*, *17*(4), pp.416-427.

[486] Ferwerda, J., 2009. The economics of crime and money laundering: does anti-money laundering policy reduce crime?. *Review of Law & Economics*, *5*(2), pp.903-929.

transactions involving high-risk jurisdictions.[487] By leveraging advanced analytics and machine learning algorithms, organizations can enhance their ability to identify potential money laundering activities and take appropriate action to mitigate risks.

In addition to proactive monitoring, AML considerations also involve conducting thorough investigations into suspicious activities and reporting them to regulatory authorities as required by law.[488] Financial institutions and businesses must maintain comprehensive records of their AML efforts and cooperate with law enforcement agencies and regulatory bodies in investigations related to money laundering or other financial crimes. By fostering a culture of compliance and accountability, organizations can strengthen their AML frameworks and contribute to the collective efforts to combat financial crime.

Anti-Terrorism Considerations

Anti-terrorism measures are of paramount importance for companies operating in the web3 or blockchain space due to the potential risks associated with terrorist financing activities.[489] Terrorist organizations have increasingly turned to innovative technologies like cryptocurrencies to raise funds, transfer money, and support their operations while evading detection by authorities.[490] As such, web3 companies must implement robust anti-terrorism protocols to prevent their platforms from being exploited for illicit purposes and to comply with regulatory requirements.

---

[487] *Id.*
[488] Kemal, *supra.*
[489] Ramraj, V.V., Hor, M. and Roach, K. eds., 2009. *Global anti-terrorism law and policy*. Cambridge University Press.
[490] Keeney, R.L., 2007. Modeling values for anti-terrorism analysis. *Risk Analysis: An International Journal*, *27*(3), pp.585-596.

One key aspect of anti-terrorism measures is the implementation of stringent Know Your Customer (KYC) procedures to verify the identities of customers and screen them against global watchlists of known terrorists and sanctioned individuals.[491] By conducting thorough due diligence on customers and monitoring their transactions, companies can identify and flag suspicious activities that may be indicative of terrorist financing attempts. Compliance with KYC requirements not only helps prevent the use of web3 platforms for terrorist financing but also ensures regulatory compliance and fosters trust among users and stakeholders.

Web3 companies must enhance their transaction monitoring capabilities to detect and disrupt potential terrorist financing activities on their platforms. This involves deploying advanced analytics and artificial intelligence algorithms to analyze transactional data for patterns and anomalies that may signal illicit behavior.[492] By proactively monitoring transactions and promptly reporting any suspicious activities to relevant authorities, companies can contribute to the global efforts to combat terrorism financing and disrupt terrorist networks' funding streams.

In addition to robust KYC and transaction monitoring procedures, web3 companies should collaborate closely with law enforcement agencies, intelligence organizations, and regulatory bodies to share information and intelligence related to terrorist financing activities.[493] Information

---

[491] Camilleri, J., 2002. Terrorism, anti-terrorism and the globalization of insecurity. *Arena Journal*, (19), pp.7-19.

[492] Macklem, P., Roach, K. and Daniels, R.J. eds., 2001. *The security of freedom: Essays on Canada's anti-terrorism bill*. University of Toronto Press.

sharing allows for timely identification and response to emerging threats, facilitates the investigation and prosecution of individuals involved in terrorist financing, and strengthens the overall resilience of the financial system against illicit actors.

Web3 companies must invest in employee training and awareness programs to educate their staff about the risks and red flags associated with terrorist financing and other financial crimes.[494] By promoting a culture of vigilance and compliance within their organizations, companies can empower their employees to identify and report suspicious activities effectively. Additionally, ongoing education and training ensure that employees remain informed about evolving threats and regulatory developments, enabling them to adapt their practices accordingly and uphold the highest standards of integrity and security.

Anti-Human Trafficking Considerations

Anti-human trafficking considerations are crucial for web3 security as the decentralized and pseudonymous nature of blockchain technology presents unique challenges and opportunities for combating this heinous crime.[495] Human trafficking, involving the exploitation and coercion of individuals for forced labor or commercial sex, remains a pervasive and abhorrent global issue, with traffickers increasingly utilizing digital platforms to facilitate their operations. In response, web3 companies must prioritize anti-human trafficking measures to prevent their platforms

[493] Stepanova, E., 2003. *Anti-terrorism and peace-building during and after conflict* (pp. 10-36). Stockholm.
[494] *Id.*
[495] Davy, D., 2016. Anti–human trafficking interventions: How do we know if they are working?. *American journal of evaluation*, *37*(4), pp.486-504.

from being abused by traffickers while safeguarding the rights and dignity of vulnerable individuals.

First and foremost, web3 companies must implement robust Know Your Customer (KYC) and customer due diligence procedures to verify the identities of users and screen them against databases of known traffickers and victims.[496] By conducting thorough identity verification and risk assessments, companies can identify and flag suspicious activities indicative of human trafficking, such as irregular transaction patterns or the use of multiple accounts to evade detection. Moreover, KYC protocols enable companies to comply with regulatory requirements aimed at preventing financial institutions from unwittingly facilitating human trafficking-related transactions.

Web3 companies should leverage the transparency and immutability of blockchain technology to enhance traceability and accountability throughout supply chains, particularly in industries susceptible to human trafficking, such as agriculture, manufacturing, and hospitality.[497] By implementing blockchain-based solutions for supply chain management and product tracking, companies can verify the provenance of goods, monitor labor conditions, and detect any instances of forced labor or exploitation along the supply chain. This increased transparency not only helps companies identify and address human trafficking

---

[496] Szablewska, N. and Kubacki, K., 2018. Anti-human trafficking campaigns: A systematic literature review. *Social marketing quarterly*, *24*(2), pp.104-122.
[497] Kim, H.W., Park, T., Quiring, S. and Barrett, D., 2018. The anti-human trafficking collaboration model and serving victims: Providers' perspectives on the impact and experience. *Journal of evidence-informed social work*, *15*(2), pp.186-203.

risks but also builds trust among consumers who prioritize ethical sourcing and corporate responsibility.

Additionally, web3 companies can leverage advanced data analytics and artificial intelligence tools to analyze transactional data and detect patterns indicative of human trafficking activities.[498] By monitoring transactional behavior and identifying suspicious patterns, companies can proactively intervene to disrupt trafficking networks, assist law enforcement agencies in their investigations, and prevent further exploitation of victims. Moreover, collaboration with law enforcement, non-governmental organizations, and other stakeholders is essential for sharing information and coordinating efforts to combat human trafficking effectively.[499]

Web3 companies should prioritize the development and adoption of responsible business practices and ethical guidelines that uphold human rights and dignity. This includes conducting due diligence on business partners and suppliers to ensure they adhere to ethical labor practices and do not engage in or facilitate human trafficking. Companies should also provide training and resources to employees to raise awareness about the signs of human trafficking and empower them to report suspicious activities confidently.[500] By fostering a culture of corporate

---

[498] McDonald, W.F., 2014. Explaining the under-performance of the anti-human-trafficking campaign: experience from the United States and Europe. *Crime, Law and Social Change*, *61*, pp.125-138.
[499] Bravo, K.E., 2010. The role of the transatlantic slave trade in contemporary anti-human trafficking discourse. *Seattle J. Soc. Just.*, *9*, p.555.
[500] Hounmenou, C., 2020. Engaging anti-human trafficking stakeholders in the research process. *Journal of Human Trafficking*, *6*(1), pp.30-49.

social responsibility and ethical conduct, web3 companies can contribute to the global fight against human trafficking and create safer, more secure online environments for all users.

Conclusion

KYC protocols are integral to web3 security, serving as a crucial line of defense against illicit activities such as money laundering, terrorism financing, and human trafficking. By implementing robust KYC procedures, web3 companies can verify the identities of their users, mitigate the risks of financial crime, and ensure compliance with regulatory requirements. However, effective KYC implementation requires a balanced approach that prioritizes security while respecting user privacy and maintaining a seamless user experience.

The adoption of advanced technologies such as blockchain, artificial intelligence, and data analytics can enhance the effectiveness of KYC processes, enabling companies to detect and prevent illicit activities more efficiently. By harnessing the transparency and traceability of blockchain technology, companies can strengthen their KYC capabilities, improve risk management, and foster greater trust and confidence among stakeholders. Additionally, collaboration with regulatory authorities, law enforcement agencies, and other industry stakeholders is essential for developing standardized KYC frameworks, sharing intelligence, and addressing emerging threats collectively.

The importance of KYC extends beyond regulatory compliance, playing a pivotal role in safeguarding the integrity of the web3 ecosystem and protecting users from harm. As web3 technologies continue to evolve and permeate various industries, the need for effective KYC measures becomes increasingly paramount to ensure the resilience and security of digital ecosystems. By embracing KYC best practices, web3 companies can uphold the highest standards of integrity, trustworthiness, and accountability, thereby fostering a safer and more resilient digital economy for all stakeholders.

Chapter 25: Sybil Attacks and Identity Fraud

"The true identity theft is not financial. It's not in cyberspace. It's spiritual. It's been taken."

Stephen Covey[501]

Chapter 25 analyzes the phenomenon of Sybil attacks and identity fraud, two tactics that pose significant threats to the integrity and security of web3 ecosystems. Sybil attacks involve the creation of multiple fake identities or nodes by a single malicious entity to manipulate or compromise a network's operations. Conversely, identity fraud pertains to the illicit use of stolen or fabricated personal information to impersonate legitimate users, deceive systems, or commit fraudulent activities. Both tactics exploit vulnerabilities in identity verification mechanisms, undermining trust, and endangering the reliability of digital transactions and interactions.

Sybil attacks represent a pervasive challenge in decentralized networks, where participants often lack centralized authority or oversight. Malicious actors exploit this decentralized structure to create numerous fake identities, enabling them to influence network behavior, manipulate voting processes, or launch coordinated attacks. By masquerading as multiple distinct entities, attackers can deceive consensus algorithms, disrupt network operations, or gain unauthorized access to resources. As such, detecting and mitigating Sybil attacks is paramount to preserving the integrity and functionality of decentralized systems.

---

[501] (2024). "Identity Theft Quotes, Stephen Covey" Brainy Quote at https://www.brainyquote.com/topics/identity-theft-quotes.

Identity fraud poses significant risks to individuals, businesses, and online platforms, as perpetrators exploit stolen or falsified personal information to assume false identities and perpetrate fraudulent activities. From financial fraud and identity theft to account takeover and social engineering scams, identity fraud encompasses a wide range of malicious behaviors that can have devastating consequences for victims. Moreover, the proliferation of digital platforms and the increasing digitization of personal data have amplified the prevalence and sophistication of identity fraud schemes, necessitating robust safeguards and countermeasures.

To mitigate the risks posed by Sybil attacks and identity fraud, web3 companies and decentralized platforms must implement comprehensive security measures and identity verification protocols. Proactive measures such as multi-factor authentication, cryptographic signatures, and reputation-based systems can help thwart Sybil attacks by enhancing the authenticity and trustworthiness of network participants. Additionally, robust identity verification processes, Know Your Customer (KYC) procedures, and data validation techniques can help prevent identity fraud by verifying the legitimacy of users and detecting suspicious activities.

Leveraging advanced technologies such as blockchain, biometrics, and machine learning can bolster the resilience and effectiveness of anti-Sybil and anti-fraud mechanisms. Blockchain-based identity solutions offer tamper-resistant and verifiable identity credentials, reducing the risk of identity manipulation or forgery. Biometric authentication methods, such as fingerprint or facial recognition, provide enhanced security and accuracy in verifying users' identities, while machine learning algorithms can analyze vast datasets to detect anomalous behavior and potential fraud indicators.

Addressing the challenges posed by Sybil attacks and identity fraud requires a multifaceted approach that combines technological innovations, robust security protocols, and proactive risk management strategies. By adopting proactive measures to authenticate users, validate identities, and detect suspicious activities, web3 companies can fortify their defenses against Sybil attacks and identity fraud, safeguarding the integrity and trustworthiness of digital ecosystems.

Avoiding Sybil Attacks

To mitigate the risks of Sybil attacks, companies operating in web3 environments must implement several proactive measures aimed at enhancing network security and integrity. First, implementing robust identity verification mechanisms is crucial for validating the authenticity of network participants and preventing the creation of fake or duplicate identities.[502] By requiring users to undergo thorough identity verification processes, such as Know Your Customer (KYC) procedures or multi-factor authentication, companies can mitigate the risk of Sybil attacks by ensuring that each participant represents a legitimate and unique entity.

Second, leveraging reputation-based systems and cryptographic signatures can help deter Sybil attacks by establishing trust among network participants and verifying the integrity of transactions or interactions.[503] Reputation-

---

[502] Douceur, J.R., 2002, March. The sybil attack. In *International workshop on peer-to-peer systems* (pp. 251-260). Berlin, Heidelberg: Springer Berlin Heidelberg.

based systems assign reputation scores to individual entities based on their past behavior and interactions within the network, allowing other participants to assess their trustworthiness and reliability. Additionally, cryptographic signatures enable users to sign and authenticate their messages or transactions using unique cryptographic keys, thereby preventing unauthorized entities from impersonating legitimate users or nodes.

Furthermore, implementing consensus algorithms and network governance mechanisms that are resilient to Sybil attacks is essential for maintaining the security and integrity of decentralized systems. Consensus algorithms such as Proof of Work (PoW) or Proof of Stake (PoS) rely on computational or stake-based mechanisms to validate transactions and secure the network against malicious actors.[504] By requiring participants to invest significant computational resources or stake a certain amount of cryptocurrency to participate in the consensus process, these algorithms make it economically infeasible for attackers to orchestrate Sybil attacks.

Lastly, fostering transparency and accountability within the network ecosystem can help mitigate the risks of Sybil attacks by enabling participants to monitor and report suspicious activities effectively. Companies should establish clear guidelines and protocols for reporting and addressing potential instances of identity fraud or malicious behavior.[505] Additionally, implementing robust monitoring and auditing mechanisms can help detect and mitigate Sybil

---

[503] *Id.*
[504] Zhang, K., Liang, X., Lu, R. and Shen, X., 2014. Sybil attacks and their defenses in the internet of things. *IEEE Internet of Things Journal*, *1*(5), pp.372-383.
[505] *Id.*

attacks in real-time, allowing companies to respond promptly and effectively to emerging threats. By promoting transparency, accountability, and vigilance within the network, companies can create a more resilient and secure environment that mitigates the risks posed by Sybil attacks.

What can companies do to avoid Identity Fraud

Preventing identity fraud within web3 environments requires the implementation of comprehensive protocols and security measures to verify the authenticity of user identities and safeguard sensitive information. First, companies can employ robust identity verification processes, such as Know Your Customer (KYC) procedures, to validate the identity of users and ensure compliance with regulatory requirements.[506] KYC protocols typically involve verifying users' identities using government-issued identification documents, biometric data, and other reliable sources of information, thereby minimizing the risk of identity fraud and unauthorized access to sensitive systems or data.

Second, implementing multi-factor authentication (MFA) mechanisms can significantly enhance the security of user accounts and mitigate the risk of identity fraud.[507] MFA requires users to provide multiple forms of verification, such as passwords, biometric data, one-time passcodes, or physical tokens, before gaining access to their accounts or performing sensitive transactions. By adding an extra layer of security beyond traditional password-based authentication, MFA protocols help prevent unauthorized

---

[506] Bambacht, J. and Pouwelse, J., 2022. Web3: A decentralized societal infrastructure for identity, trust, money, and data. *arXiv preprint arXiv:2203.00398.*
[507] *Id.*

access to user accounts and reduce the likelihood of identity theft or impersonation.

Deploying advanced encryption technologies, such as public-key cryptography and digital signatures, can help protect user identities and sensitive data from unauthorized access or tampering.[508] Encryption protocols encrypt data during transmission and storage, making it unreadable to unauthorized parties even if intercepted or compromised. Digital signatures provide a means of verifying the authenticity and integrity of digital documents or transactions, ensuring that they have not been altered or manipulated by malicious actors.

Leveraging blockchain technology and decentralized identity solutions can enhance the security and trustworthiness of user identities within web3 environments. Blockchain-based identity platforms enable users to maintain control over their personal data and selectively disclose information to trusted parties, reducing the risk of identity fraud and data breaches.[509] Decentralized identity solutions use blockchain-based identifiers and cryptographic techniques to create self-sovereign identities that are resistant to tampering or manipulation by centralized authorities or malicious actors.

Implementing comprehensive fraud detection and prevention systems can help companies proactively identify

---

[508] DeDominicis, B.E., 2022. Web3, Hegemony And Anonymity: The Transnational Social Identity Dynamics Of Globalized National Political Economic Interdependency. *Global Journal of Business Research*, *16*(1), pp.91-119.

[509] Winter, P., Lorimer, A.H., Snyder, P. and Livshits, B., 2021. Security, Privacy, and Decentralization in Web3. *arXiv preprint arXiv:2109.06836.*

and mitigate potential instances of identity fraud before they escalate into more significant security incidents. These systems leverage machine learning algorithms, anomaly detection techniques, and behavioral analytics to detect suspicious patterns or activities indicative of fraudulent behavior, allowing companies to take timely action to prevent unauthorized access or mitigate potential risks to user security and privacy.[510] By adopting a proactive approach to identity fraud prevention and investing in robust security measures and protocols, companies can effectively protect user identities and maintain the integrity of their web3 ecosystems.

Conclusion

Mitigating the risks associated with Sybil attacks and identity fraud is paramount for the security and integrity of web3 environments. Sybil attacks, characterized by the creation of multiple fake identities to manipulate or compromise systems, pose significant challenges to the trustworthiness and reliability of online platforms. Likewise, identity fraud, involving the unauthorized use of individuals' personal information for fraudulent purposes, presents serious threats to user privacy and security.

To address these challenges, companies must implement robust security measures and protocols designed to prevent and detect instances of Sybil attacks and identity fraud effectively. This includes employing stringent identity verification processes, such as Know Your Customer (KYC) procedures, to authenticate user identities and validate the accuracy of provided information. The deployment of multi-factor authentication (MFA) mechanisms and encryption technologies can enhance the

510

security of user accounts and data, reducing the risk of unauthorized access and data breaches.

Furthermore, leveraging blockchain technology and decentralized identity solutions can provide more secure and trustworthy methods for managing user identities and mitigating the risks of Sybil attacks and identity fraud. Blockchain-based identity platforms offer enhanced transparency, immutability, and tamper resistance, making them ideal for establishing verifiable and tamper-proof digital identities. By empowering users to maintain control over their personal data and selectively disclose information to trusted parties, decentralized identity solutions can help prevent identity fraud and protect user privacy.

Proactive monitoring and detection of suspicious activities through fraud detection and prevention systems are essential for identifying and mitigating potential instances of Sybil attacks and identity fraud before they escalate. By leveraging advanced analytics, machine learning algorithms, and anomaly detection techniques, companies can detect and respond to abnormal behavior indicative of fraudulent activity, safeguarding user accounts and maintaining the integrity of their web3 ecosystems.

Combating Sybil attacks and identity fraud requires a multifaceted approach that combines robust security measures, innovative technologies, and proactive monitoring strategies. By prioritizing the security and trustworthiness of user identities within web3 environments, companies can foster a safer and more secure digital ecosystem for all stakeholders involved.

Chapter 26: Accessibility of Digital Identity

"Now I existed solely thanks to the quantum paradox, my brain a collection of qubits in quantum superposition, encoding truths and memories, imagination and irrationality in opposing, contradictory states that existed and didn't exist, all at the same time."

Robin Wasserman, Crashed[511]

In the digital age, accessibility of digital identity has become increasingly important, balancing the need for convenience with robust security measures. This chapter explores into the intricate balance between ease of use and security, emphasizing the importance of empowering users while enhancing their security and ensuring interoperability between systems for a seamless experience. By prioritizing these aspects, companies can create a user-centric approach to digital identity management that fosters trust, enhances security, and promotes usability.

Empowering users and fostering ownership of digital identity lie at the core of creating a secure and user-centric digital ecosystem. By giving users greater control over their digital identities, including the ability to manage and authenticate their identities across various platforms and services, companies can empower individuals to protect their privacy and secure their online presence. This involves implementing user-friendly tools and interfaces that facilitate identity management tasks, such as account creation, authentication, and consent management, while also providing transparency and control over the use of personal data.

---

[511] (2024). "Digital Identity Quotes, Robin Wasserman." Good Reads at https://www.goodreads.com/quotes/tag/digital-identity.

Enhancing the security of users' digital identities is paramount to mitigating risks and safeguarding against unauthorized access and identity theft. This entails implementing robust security measures, such as multi-factor authentication (MFA), encryption, and biometric authentication, to strengthen the authentication process and protect user credentials from compromise. Additionally, adopting decentralized identity solutions built on blockchain technology can offer enhanced security, privacy, and user control by eliminating single points of failure and enabling self-sovereign identity management.

Interoperability between systems is essential for ensuring seamless user experiences across different platforms and services while maintaining security and privacy standards. By adopting open standards and protocols for digital identity management, companies can facilitate the interoperability of identity systems, allowing users to seamlessly access and authenticate their identities across diverse ecosystems. This interoperability fosters collaboration and innovation while ensuring that users can securely interact with various services and applications without compromising their privacy or security.

Furthermore, promoting a user-centric approach to digital identity management involves prioritizing usability and accessibility for all individuals, including those with disabilities or limited technical expertise. Companies should strive to design inclusive and user-friendly interfaces that accommodate diverse user needs and preferences, ensuring that everyone can easily navigate and interact with digital identity systems. By prioritizing accessibility and usability, companies can enhance user adoption and satisfaction while promoting digital inclusion and equity.

The accessibility of digital identity hinges on empowering users, enhancing security, and promoting interoperability between systems for a seamless and secure user experience. By prioritizing these aspects and adopting user-centric approaches to digital identity management, companies can build trust, foster innovation, and promote inclusivity in the digital realm. Through collaboration, innovation, and a commitment to user privacy and security, we can create a more accessible and secure digital future for all.

Empowering Users and Ownership of Their Digital Identity

Empowering users with control over their digital identities is crucial for maintaining security within a company. By instilling a sense of ownership and responsibility for their identities, employees are more likely to prioritize security practices and protect their personal and professional information from unauthorized access or exploitation.[512] Digital identity management extends beyond the confines of the workplace, encompassing various online activities such as social media interactions, web browsing habits, and personal communications. Recognizing the interconnected nature of digital identities, employees must understand the implications of their online presence on both personal and professional fronts.

Empowering users with ownership of their digital identities fosters a culture of accountability and proactive security measures within the company.[513] When employees feel a

[512] Takemiya, M. and Vanieiev, B., 2018, July. Sora identity: Secure, digital identity on the blockchain. In *2018 ieee 42nd annual computer software and applications conference (compsac)* (Vol. 2, pp. 582-587). IEEE

sense of ownership over their digital personas, they are more likely to take precautions to safeguard their online activities and personal data. This includes implementing strong authentication methods, regularly updating passwords, and being vigilant against phishing attempts or social engineering attacks that may compromise their identities. By empowering users to actively manage and protect their digital identities, companies can strengthen their overall security posture and mitigate the risk of data breaches or insider threats.

Recognizing the potential security risks associated with employees' digital identities outside of the workplace is paramount for safeguarding sensitive company information. Social media platforms, online forums, and other digital channels can serve as avenues for cybercriminals to gather intelligence or exploit vulnerabilities within the organization.[514] Employees may inadvertently expose themselves to threats such as identity theft, blackmail, or phishing scams, which could ultimately compromise company data or systems. Therefore, educating employees about the importance of maintaining a secure online presence and the potential repercussions of compromising their digital identities is essential for mitigating security risks.

Incorporating digital identity management practices into employee training and awareness programs can help reinforce security best practices and promote a culture of vigilance and responsibility.[515] By providing employees

---

[513] Windley, P.J., 2005. *Digital Identity: Unmasking identity management architecture (IMA).* " O'Reilly Media, Inc.".
[514] Hamer, T., Taylor, K., Ng, K.S. and Tiu, A., 2019. Private digital identity on blockchain.
[515] Takemiya, *supra.*

with the knowledge and tools necessary to protect their digital identities, companies can empower them to make informed decisions and mitigate potential security threats effectively. This includes educating employees about the risks associated with oversharing personal information online, the importance of using privacy settings and security features on social media platforms, and the need to exercise caution when interacting with unknown or suspicious entities online.

Ultimately, by empowering employees with ownership of their digital identities and promoting security-conscious behavior both inside and outside the workplace, companies can enhance their overall security posture and minimize the risk of data breaches or cyberattacks.[516] By fostering a culture of security awareness and accountability, organizations can better protect their assets, maintain trust with customers and partners, and navigate the evolving threat landscape in today's digital age.

Enhancing the Security and Privacy of Users

Enhancing the security of a company's digital identity while respecting users' privacy requires a balanced approach that incorporates robust security measures without compromising individual privacy rights. One methodology is the implementation of multi-factor authentication (MFA) systems, which provide an additional layer of security beyond traditional password-based authentication methods.[517] By requiring users to authenticate using multiple factors such as passwords,

[516] *Id.*

[517] Guo, S., Zhang, F., Guo, S., Xu, S. and Qi, F., 2023. Blockchain-assisted privacy-preserving data computing architecture for Web3. *IEEE Communications Magazine*, *61*(8), pp.28-34.

biometrics, or security tokens, companies can significantly reduce the risk of unauthorized access while still respecting users' privacy by not relying solely on sensitive personal information.

Another effective strategy is the adoption of encryption technologies to protect sensitive data and communications. End-to-end encryption ensures that data remains encrypted throughout transmission, storage, and processing, thereby safeguarding it from unauthorized interception or access. By encrypting sensitive communications and data stored on company servers, organizations can enhance security while preserving users' privacy by ensuring that only authorized parties have access to the encrypted information.

Implementing robust access control mechanisms is also essential for maintaining security and privacy within the company's digital ecosystem. Role-based access control (RBAC) systems allow organizations to define and enforce granular access permissions based on users' roles and responsibilities, ensuring that individuals only have access to the resources and information necessary for their job functions.[518] By carefully managing access privileges and monitoring user activity, companies can mitigate the risk of unauthorized access while still respecting users' privacy rights by limiting access to sensitive data and systems.

Additionally, companies can leverage anonymization techniques to protect users' privacy while still obtaining valuable insights from data analytics and business intelligence initiatives. Anonymizing personally identifiable information (PII) before processing or storing it

---

[518] Ma, W. and Huang, K., 2022. *Blockchain and Web3: Building the cryptocurrency, privacy, and security foundations of the metaverse.* John Wiley & Sons.

helps prevent the unauthorized tracking or profiling of individual users while still allowing companies to derive useful insights from aggregated data sets. By anonymizing data and implementing strict data protection policies, organizations can enhance security and privacy simultaneously, ensuring compliance with privacy regulations and safeguarding users' personal information from unauthorized access or misuse.

Furthermore, transparency and user consent are critical aspects of maintaining privacy and security within a company's digital identity framework. Clearly communicating privacy policies, data handling practices, and security measures to users helps build trust and confidence while empowering individuals to make informed decisions about their digital interactions. Providing users with control over their personal data through robust privacy settings and consent mechanisms allows them to manage their privacy preferences effectively and ensures that their rights are respected throughout the digital identity lifecycle.

Finally, ongoing monitoring, auditing, and compliance assessments are essential for ensuring the effectiveness of security and privacy measures within the company's digital identity ecosystem.[519] Regularly reviewing access logs, conducting security audits, and assessing compliance with relevant regulations help identify and address potential security vulnerabilities or privacy risks proactively. By continuously monitoring and improving security and privacy practices, organizations can maintain a strong

---

[519] Bambacht, J. and Pouwelse, J., 2022. Web3: A decentralized societal infrastructure for identity, trust, money, and data. *arXiv preprint arXiv:2203.00398.*

security posture while upholding users' privacy rights and trust in the digital environment.

Ensuring that there is interoperability between systems and a seamless experience.

Ensuring interoperability between systems within a company is crucial for streamlining workflows, enhancing productivity, and maintaining robust security measures. When employees are required to use multiple logins across several systems, it not only leads to inefficiencies but also increases the risk of security vulnerabilities. By integrating systems and implementing a unified authentication framework, organizations can simplify access management while bolstering security measures.

One of the primary benefits of ensuring interoperability between systems is the elimination of the need for employees to manage multiple passwords and login credentials.[520] With a unified authentication system, users can access various applications and resources using a single set of credentials, reducing the complexity of password management and minimizing the likelihood of password-related security incidents such as credential theft or phishing attacks. By centralizing authentication processes, organizations can enhance security while improving user experience and productivity.

Interoperability enables organizations to implement secondary authentication mechanisms within the system, such as multi-factor authentication (MFA), without requiring users to use separate devices or login procedures.

---

[520] Yildiz, H., Küpper, A., Thatmann, D., Göndör, S. and Herbke, P., 2022. A tutorial on the interoperability of self-sovereign identities. *arXiv preprint arXiv:2208.04692.*

MFA adds an extra layer of security by requiring users to provide additional authentication factors beyond passwords, such as biometrics or security tokens. By integrating MFA capabilities directly into the authentication system, organizations can enforce stronger security measures without inconveniencing users with additional login steps or devices.

Another advantage of interoperability is the ability to enforce access controls and permissions consistently across all integrated systems and applications. By centrally managing access policies and permissions, organizations can ensure that users only have access to the resources and information relevant to their roles and responsibilities, reducing the risk of unauthorized access or data breaches.[521] By integrating access control mechanisms into the authentication system, organizations can enforce granular access controls while maintaining a seamless user experience.

Interoperability also facilitates the implementation of robust identity and access management (IAM) solutions, which provide centralized control over user identities, access rights, and authentication processes. IAM solutions enable organizations to automate user provisioning and deprovisioning, enforce strong authentication policies, and monitor user activity across integrated systems. By integrating IAM functionalities into the authentication framework, organizations can enhance security while simplifying identity management and access control processes.

---

[521] Palfrey, J. and Gasser, U., 2009. Digital identity interoperability and einnovation. *Berkman Publication Series.*

Furthermore, interoperability enables organizations to leverage advanced security features such as single sign-on (SSO) capabilities, which allow users to authenticate once and access multiple applications and systems seamlessly. SSO enhances user convenience and productivity while reducing the risk of password-related security incidents. By integrating SSO functionalities into the authentication system, organizations can provide users with frictionless access to resources while maintaining stringent security measures.

In addition to security benefits, interoperability also contributes to operational efficiency and cost-effectiveness by reducing the complexity of managing disparate systems and applications. By standardizing authentication processes and integrating systems, organizations can streamline administrative tasks, reduce support overhead, and lower operational costs associated with managing multiple authentication solutions. Interoperability enables organizations to leverage existing infrastructure and investments more effectively while adapting to evolving security requirements and business needs.

Overall, ensuring interoperability between systems within a company is essential for enhancing security, improving user experience, and optimizing operational efficiency. By integrating authentication, access control, and identity management functionalities into a unified framework, organizations can enforce robust security measures while providing users with seamless access to resources across various systems and applications. Interoperability enables organizations to strike a balance between security and usability, ensuring that employees can access the tools and information they need to perform their roles effectively while maintaining stringent security standards.

Conclusion

Ensuring the accessibility of digital identity while maintaining robust security measures is essential for modern organizations operating in the web3 environment. The battle between ease of use and security presents a significant challenge, but it is imperative for security to prevail in safeguarding sensitive data and protecting against cyber threats. By empowering users and fostering a sense of ownership of their digital identities, organizations can promote a culture of security awareness and responsibility among employees.

Enhancing the security of users within a company's digital ecosystem requires a multifaceted approach that balances privacy, usability, and protection against threats. By implementing advanced authentication mechanisms, access controls, and identity management solutions, organizations can strengthen security while preserving user privacy and convenience. Interoperability between systems plays a crucial role in streamlining authentication processes, enforcing access controls, and ensuring consistent security measures across all integrated systems and applications.

Integrating security protocols such as multi-factor authentication (MFA), single sign-on (SSO), and identity and access management (IAM) solutions into the authentication framework enables organizations to enforce stringent security measures without compromising user experience. These technologies not only enhance security but also streamline user access to resources, reducing friction and improving productivity. Additionally, interoperability facilitates the implementation of robust security protocols while minimizing the complexity of managing disparate systems and applications.

As organizations continue to evolve and adapt to the challenges of the digital landscape, it is essential to prioritize security without sacrificing usability. By investing in advanced security technologies, implementing best practices, and fostering a culture of security awareness, organizations can mitigate risks, protect sensitive data, and maintain trust with customers and stakeholders. Ultimately, the accessibility of digital identity should be viewed as a foundational element of modern business operations, supported by robust security measures that safeguard both user privacy and organizational assets.

Conclusion Part V

In conclusion, Part V of this comprehensive guide has delved into various aspects of security in the web3 environment, addressing critical challenges and proposing solutions to mitigate risks. From exploring the nuances of anonymity and digital identity to navigating the complexities of authentication and access control, each chapter has offered valuable insights into safeguarding sensitive data and protecting against cyber threats.

Chapter 23 highlighted the unique challenges posed by anon culture in web3 and the difficulties of implementing Know Your Customer (KYC) protocols in anonymous environments. Despite the inherent complexities, establishing robust security measures and maintaining a balance between anonymity and accountability is crucial for ensuring trust and integrity in web3 ecosystems.

Chapter 24 delved into the importance of KYC protocols in combating money laundering, terrorism financing, and human trafficking. By adhering to stringent regulatory requirements and implementing effective KYC procedures, organizations can mitigate legal and reputational risks while contributing to global efforts to combat illicit activities.

Chapter 25 explored the dangers of Sybil attacks and identity fraud, emphasizing the need for proactive measures to prevent fraudulent activities and protect digital identities. By implementing robust authentication mechanisms and leveraging advanced fraud detection technologies, organizations can enhance security and safeguard sensitive information from unauthorized access.

Chapter 26 focused on enhancing the accessibility of digital identity while maintaining strong security measures. By empowering users and promoting a culture of security awareness, organizations can strike a balance between usability and protection, fostering trust and confidence in their digital ecosystems.

Moving forward, it is imperative for organizations to remain vigilant and proactive in addressing emerging threats and evolving security challenges. By staying abreast of the latest developments in cybersecurity, adopting best practices, and leveraging advanced technologies, organizations can effectively mitigate risks and protect their digital assets.

Furthermore, collaboration and information sharing among industry stakeholders are essential for promoting collective resilience and strengthening cybersecurity posture. By fostering partnerships and collaborating on threat intelligence sharing initiatives, organizations can enhance their ability to detect, respond to, and mitigate cyber threats effectively.

In conclusion, while the web3 landscape presents numerous opportunities for innovation and growth, it also brings inherent risks and challenges. By embracing a proactive and holistic approach to cybersecurity, organizations can navigate the complexities of the digital landscape with confidence, ensuring the integrity, confidentiality, and availability of their data assets.

Part VI: Algorithmic Bias and Discrimination

Introduction to Part VI

Part VI delves into the subject of algorithmic bias and censorship, shedding light on the profound implications these phenomena have on corporate security and ethical conduct. In an era dominated by digital technologies and data-driven decision-making, the prevalence of algorithmic bias poses significant challenges for organizations seeking to maintain fairness, transparency, and integrity in their operations. The management of censorship within corporate communications adds another layer of complexity, raising critical questions about the balance between information control and freedom of expression. This part explores these issues in depth, offering insights into the mechanisms, consequences, and management strategies associated with algorithmic bias and censorship within corporate contexts.

Chapter 27 examines algorithmic bias, a pervasive phenomenon characterized by the presence of discriminatory patterns or outcomes in automated decision-making systems. From biased algorithms perpetuating systemic inequalities to the erosion of privacy and susceptibility to malicious exploitation, algorithmic bias poses multifaceted risks to corporate security and ethical governance. By unpacking the root causes and implications of algorithmic bias, this chapter aims to equip organizations with the knowledge and tools needed to identify, mitigate, and address bias in their technological systems and processes.

In Chapter 28, the focus shifts to systematic bias, exploring the broader socio-cultural dynamics that shape corporate environments and influence security practices. Institutional bias, cultural prejudice, discrimination in programming, and economic disparities can all contribute to systemic biases that undermine the fairness, inclusivity, and effectiveness of security measures within organizations. By fostering diversity, equity, and inclusion initiatives and promoting ethical leadership and decision-making, companies can mitigate the impact of systematic bias and cultivate a more resilient and socially responsible security culture.

Chapter 29 breaks into the complex terrain of censorship management within corporate communications, where the tension between information control and transparency comes to the forefront. While censorship may be necessary to maintain a unified message and protect sensitive information, excessive or arbitrary censorship can stifle creativity, innovation, and critical discourse, ultimately undermining trust and credibility. This chapter explores the challenges and opportunities associated with censorship management, offering guidance on striking the right balance between information control and freedom of expression to uphold ethical standards and foster a culture of open communication and accountability.

As organizations navigate the intricate interplay between algorithmic bias, systematic bias, and censorship within corporate contexts, it becomes increasingly clear that a nuanced and holistic approach is needed to address these challenges effectively. By integrating principles of fairness, transparency, and accountability into their security practices and corporate culture, companies can build trust, resilience, and legitimacy in an era defined by rapid technological change and socio-cultural complexity. This part endeavors to provide practical insights, strategies, and best practices to empower organizations to navigate the complexities of algorithmic bias and censorship while upholding their ethical commitments and security imperatives.

Chapter 27: Algorithmic Bias

"Modernity has become algorithms. Our reality is now intercepted by the exploration and exploitation of our psychic cues that rewrite history by rerouting consumption, elections, public opinions and civil war."

Roger Spitz, The Definitive Guide to Thriving on Disruption[522]

Algorithmic bias poses a significant threat to organizational security by perpetuating discrimination and unequal access to resources and opportunities. As algorithms increasingly dictate decision-making processes across various domains, including hiring, lending, and policing, the propagation of biased outcomes can exacerbate existing inequalities and marginalize vulnerable populations.[523] In doing so, biased algorithms not only compromise the ethical integrity of organizational practices but also fuel societal tensions and erode trust in institutional systems.

Algorithmic bias contributes to privacy erosion and data exploitation, amplifying the risks of unauthorized access and misuse of sensitive information. By perpetuating discriminatory practices and reinforcing social biases, biased algorithms can inadvertently expose individuals to heightened surveillance and exploitation, exacerbating privacy concerns and undermining the fundamental principles of data protection and confidentiality.[524] This

---

[522] (2004). "Algorithm Quotes, Roger Spitz." Good Reads at https://www.goodreads.com/quotes/tag/algorithms.

[523] Simon, J., Wong, P.H. and Rieder, G., 2020. Algorithmic bias and the Value Sensitive Design approach. *Internet Policy Review*, *9*(4).

[524] Chakraborty, J., Majumder, S. and Menzies, T., 2021, August. Bias in machine learning software: Why? how? what to do?. In *Proceedings of the 29th ACM joint meeting on European software engineering conference and symposium on the foundations of software engineering*

erosion of privacy not only compromises individual autonomy and dignity but also exposes organizations to regulatory scrutiny and legal liabilities.

Furthermore, biased algorithms create vulnerabilities to cyberattacks and malicious manipulation, as adversaries exploit inherent biases to subvert automated systems and gain unauthorized access to critical assets.[525] By exploiting vulnerabilities in algorithmic decision-making processes, attackers can manipulate outcomes, disrupt operations, and compromise the integrity and availability of organizational data, posing significant security risks and financial consequences.

Additionally, the proliferation of algorithmic bias undermines the trust, legitimacy, and loyalty of employees, customers, and stakeholders, as individuals become increasingly wary of automated systems that perpetuate unfair and discriminatory outcomes. The erosion of trust in organizational practices and decision-making processes can have far-reaching implications, affecting employee morale, customer satisfaction, and organizational reputation, ultimately jeopardizing long-term success and sustainability.

In light of these multifaceted challenges, it is imperative for organizations to address algorithmic bias proactively, implementing robust mitigation strategies and accountability mechanisms to safeguard against biased outcomes and uphold principles of fairness, equity, and transparency.[526] By fostering a culture of diversity,

---

(pp. 429-440).

[525] *Id.*

[526] Chakraborty, J., Majumder, S. and Menzies, T., 2021, August. Bias in machine learning software: Why? how? what to do?. In *Proceedings of the 29th ACM joint meeting on European software engineering*

inclusion, and ethical decision-making, organizations can mitigate the risks of algorithmic bias and cultivate trust, legitimacy, and loyalty among employees, customers, and stakeholders, thereby enhancing organizational resilience and promoting sustainable growth in the digital age.

Discrimination and Unequal Access

Discrimination within algorithms and unequal access to resources present profound security risks for companies employing web3 systems, manifesting in various dimensions that undermine organizational integrity and resilience. First, biased algorithms perpetuate discriminatory practices by amplifying existing social biases and inequalities, leading to adverse outcomes for marginalized groups and exacerbating societal tensions. In the context of web3 systems, discriminatory algorithms can result in unequal access to opportunities, services, and resources, creating disparities that compromise organizational values of fairness and equity.

Furthermore, the propagation of discriminatory algorithms poses legal and regulatory risks for companies, as they may face litigation, fines, and reputational damage for violating anti-discrimination laws and regulations.[527] Discriminatory practices not only undermine organizational compliance with legal frameworks but also erode trust and confidence among stakeholders, including employees, customers, and investors, thereby jeopardizing organizational legitimacy and viability in the marketplace.

---

conference and symposium on the foundations of software engineering (pp. 429-440).
[527] *Id.*

Unequal access to resources driven by algorithmic bias can foster resentment and disenchantment among employees and customers, leading to diminished morale, loyalty, and engagement with the organization.[528] In the context of web3 systems, where digital interactions and transactions are increasingly mediated by algorithms, disparities in access can result in exclusionary practices that alienate individuals and communities, eroding social cohesion and undermining organizational cohesion and resilience.

From a cybersecurity standpoint, discrimination within algorithms and unequal access to resources can create vulnerabilities and exploitable loopholes that adversaries may exploit to gain unauthorized access to sensitive information or disrupt organizational operations. Biased algorithms may inadvertently leak confidential data or expose vulnerabilities that malicious actors can exploit to launch cyberattacks, compromise system integrity, and undermine organizational security posture, thereby posing significant risks to data confidentiality, integrity, and availability.

Furthermore, discrimination within algorithms and unequal access to resources can undermine employee trust and confidence in organizational practices and decision-making processes, leading to decreased productivity, innovation, and collaboration within the workforce.[529] Employees who perceive bias or unfair treatment may become disengaged, disillusioned, or even adversarial towards the organization, leading to increased turnover rates, talent shortages, and diminished organizational resilience in the face of evolving threats and challenges.

---

[528] Simon et al, *supra.*
[529] *Id.*

Addressing discrimination within algorithms and promoting equal access to resources requires a multi-faceted approach that encompasses algorithmic transparency, accountability mechanisms, diversity and inclusion initiatives, and ongoing monitoring and evaluation of algorithmic systems. By fostering a culture of fairness, equity, and transparency, organizations can mitigate the security risks associated with algorithmic bias and promote a more inclusive and resilient digital ecosystem that benefits all stakeholders.

Privacy Erosion and Data Exploitation

Privacy erosion and data exploitation present significant security challenges for both companies and their customers, posing threats to individual autonomy, confidentiality, and trust in digital interactions. Firstly, the erosion of privacy through pervasive data collection and surveillance practices undermines individuals' ability to control their personal information, exposing them to risks of identity theft, fraud, and unauthorized access to sensitive data.[530] In the context of web3 systems, where data is increasingly commodified and monetized, privacy erosion can lead to the indiscriminate sharing and exploitation of personal data, compromising individuals' privacy rights and exposing them to potential harm.

Data exploitation exacerbates security vulnerabilities by enabling malicious actors to leverage stolen or compromised data for nefarious purposes, such as identity fraud, phishing attacks, and social engineering scams. As companies amass vast repositories of consumer data for profiling and targeting purposes, they become lucrative targets for cybercriminals seeking to exploit vulnerabilities

---

[530] Charaborty et al, *supra.*

in data storage, transmission, and processing systems to steal valuable information or manipulate individuals for financial gain or malicious intent.

Furthermore, privacy erosion and data exploitation erode trust and confidence in companies' ability to safeguard customer data and protect their privacy rights, leading to reputational damage, loss of customer loyalty, and regulatory scrutiny.[531] In the wake of high-profile data breaches and privacy scandals, consumers are increasingly wary of sharing their personal information with companies, fearing that their data may be misused or compromised without their consent. This erosion of trust undermines companies' credibility and legitimacy in the eyes of consumers, impeding their ability to attract and retain customers and driving them to seek alternative providers who prioritize privacy and data protection.

From a regulatory perspective, privacy erosion and data exploitation expose companies to legal and compliance risks, as they may face penalties, fines, and legal liabilities for violating data protection laws and regulations. Governments and regulatory authorities are increasingly imposing stringent requirements and standards for data handling, storage, and processing, holding companies accountable for ensuring the confidentiality, integrity, and availability of customer data and respecting individuals' privacy rights. Failure to comply with regulatory mandates can result in significant financial and reputational consequences for companies, jeopardizing their viability and sustainability in the marketplace.

---

[531] *Id.*

To mitigate the security risks associated with privacy erosion and data exploitation, companies must adopt robust data protection measures, including encryption, anonymization, access controls, and user consent mechanisms, to safeguard customer data and preserve individual privacy rights.[532] By prioritizing data security and privacy as core principles of their business operations, companies can build trust, foster customer loyalty, and differentiate themselves in an increasingly competitive and privacy-conscious market landscape.

Vulnerability to Attacks

Algorithmic bias can introduce vulnerabilities that malicious actors can exploit to perpetrate attacks on systems and individuals. When algorithms exhibit biases, such as favoring certain demographic groups or making decisions based on discriminatory patterns, they create opportunities for exploitation by adversaries seeking to manipulate or subvert their outcomes. For example, if an algorithm used in a hiring process exhibits bias against candidates from certain backgrounds, attackers could exploit this bias to game the system by artificially inflating the qualifications of preferred candidates or sabotaging the prospects of competitors.[533]

Algorithmic bias can undermine the accuracy and reliability of automated decision-making processes, leading to erroneous outcomes that could be exploited by attackers to gain unauthorized access to resources or privileges. For instance, if a facial recognition algorithm exhibits racial bias, misidentifying individuals based on their skin color, attackers could exploit this vulnerability to evade detection

---

[532] Chakraborty, *supra.*
[533] *Id.*

or impersonate legitimate users in security-sensitive applications such as access control systems or surveillance systems.

Additionally, algorithmic bias can exacerbate existing disparities and inequities in society, amplifying social tensions and grievances that could be exploited by threat actors to sow discord, incite violence, or undermine social cohesion. For example, if an algorithm used in predictive policing exhibits bias against certain communities, over-policing and discriminatory law enforcement practices could provoke backlash and civil unrest, creating opportunities for exploitation by extremist groups or hostile actors seeking to exploit social divisions for their own gain.

Furthermore, the opacity and lack of transparency surrounding algorithmic decision-making processes can obscure the presence of bias and vulnerabilities, making it difficult for organizations to detect and mitigate them effectively.[534] Attackers may exploit this lack of accountability and oversight to manipulate algorithmic systems for malicious purposes, exploiting their vulnerabilities to gain unfair advantages or perpetrate harm without fear of detection or repercussion. As a result, addressing algorithmic bias requires greater transparency, accountability, and scrutiny of algorithmic systems to identify and mitigate vulnerabilities that could be exploited by adversaries.

Erosion of Trust and Legitimacy

Problems with algorithmic bias can erode trust and legitimacy within a system, leading to significant security implications for both the company and its customers.[535]

---

[534] *Id.*

When users perceive that algorithmic systems exhibit unfairness or discrimination, it undermines their confidence in the integrity and reliability of those systems, creating doubts about the company's commitment to fairness and equity. This loss of trust can have far-reaching consequences, as users may become hesitant to engage with the system or may actively seek alternatives, thereby reducing the system's effectiveness and potentially harming the company's reputation.

The failure of machine learning algorithms to adequately address issues of bias and discrimination can lead to widespread skepticism and skepticism about the legitimacy of algorithmic decision-making processes. When users perceive that algorithmic systems are making decisions based on biased or flawed criteria, it raises questions about the fairness and impartiality of those decisions, undermining their confidence in the system's ability to act in their best interests.[536] This lack of trust can create barriers to adoption and usage, impeding the system's effectiveness and hindering its ability to deliver value to both the company and its customers.

Furthermore, algorithmic bias can exacerbate existing disparities and inequalities, reinforcing systemic biases and perpetuating cycles of discrimination and exclusion. When users experience discrimination or marginalization as a result of biased algorithmic systems, it breeds resentment and disillusionment, undermining their trust in the company and its commitment to fairness and equality.[537] This loss of trust can manifest in various forms, from reduced

---

[535] Simon et al, *supra*.
[536] *Id.*
[537] Chackraborty, *id.*

engagement with the system to outright opposition or resistance, posing significant challenges to the company's ability to achieve its objectives and maintain its competitive edge in the marketplace.

Additionally, the failure of algorithmic systems to address issues of bias and discrimination can have legal and regulatory implications, exposing the company to potential liability and reputational damage. As governments and regulatory bodies increasingly scrutinize algorithmic decision-making processes for fairness and transparency, companies that fail to address algorithmic bias effectively may face fines, lawsuits, and other penalties, as well as damage to their brand and reputation.[538] This creates a powerful incentive for companies to proactively identify and mitigate algorithmic bias to avoid legal and regulatory consequences and maintain the trust and confidence of their customers.

The lack of trust and legitimacy in algorithmic systems can undermine the company's ability to leverage data-driven insights and innovations effectively, hampering its competitiveness and inhibiting its growth potential. When users lack confidence in the fairness and accuracy of algorithmic systems, they may be reluctant to share their data or participate in data-driven initiatives, depriving the company of valuable resources and insights. This can impede the company's ability to innovate, adapt to changing market conditions, and deliver value to its customers, ultimately undermining its long-term success and sustainability.

---

538 *Id.*

Conclusion

Algorithmic bias poses significant security risks for companies employing web3 systems, manifesting in various forms such as discrimination, unequal access, privacy erosion, vulnerability to attacks, and erosion of trust and legitimacy. Addressing these challenges requires a multifaceted approach that prioritizes fairness, transparency, and accountability in algorithmic decision-making processes.

First, companies must recognize the importance of mitigating algorithmic bias to uphold principles of fairness and equity in their operations. By proactively identifying and addressing biases within their algorithms, companies can ensure that their systems treat all users fairly and impartially, enhancing trust and confidence in their services.

Second, efforts to combat algorithmic bias should be complemented by measures to promote equal access and opportunities for all users. Companies must strive to eliminate barriers and disparities that may prevent certain individuals or groups from fully participating in their systems, thereby fostering inclusivity and diversity in their user base.

Safeguarding user privacy and data protection is paramount to maintaining trust and loyalty in web3 systems. Companies must implement robust data governance frameworks and security measures to protect sensitive information from exploitation and misuse, thereby safeguarding user privacy and upholding their rights and expectations.

Companies must remain vigilant against potential vulnerabilities and threats arising from algorithmic bias, ensuring that their systems are resilient to attacks and manipulation attempts. By investing in robust security measures and threat detection capabilities, companies can mitigate the risk of malicious exploitation and maintain the integrity and reliability of their services.

Addressing algorithmic bias requires a concerted effort from companies, regulators, and other stakeholders to develop and enforce standards and guidelines that promote fairness, transparency, and accountability in algorithmic decision-making processes. By prioritizing these principles and working collaboratively to address algorithmic bias, companies can build more secure and trustworthy web3 systems that benefit all users and stakeholders alike.

Chapter 28: Systematic Discrimination

"One of the most pervasive political [lies] of our time is the vision of liberals as compassionate and conservatives as less caring."[539]

Systematic bias can intertwine with security issues within a company employing web3 systems. This encompasses an exploration of institutional bias, cultural prejudice, discrimination inherent in programming, and the ramifications of economic disparity. These forms of bias pose significant challenges to the integrity, fairness, and effectiveness of a company's operations, potentially leading to adverse consequences for both the organization and its stakeholders.

Institutional bias refers to the systematic and often ingrained biases present within the structures and practices of organizations. This can manifest in various forms, such as unequal hiring practices, biased decision-making processes, and disparities in resource allocation. When unchecked, institutional bias can perpetuate inequality and exclusion within the company, undermining trust and confidence among employees and stakeholders.

Cultural prejudice represents another dimension of bias that can permeate the fabric of an organization, influencing its policies, norms, and interactions. This form of bias may stem from entrenched cultural beliefs, societal norms, or implicit biases held by individuals within the organization. Addressing cultural prejudice requires a concerted effort to foster diversity, equity, and inclusion, creating an

[539] (2024). "Pervasive Quotes, Thomas Sowell." Brainy Quote at https://www.brainyquote.com/topics/pervasive-quotes.

environment where all employees feel valued and respected.

Discrimination within programming highlights the inherent biases that may be present in the design, development, and implementation of algorithms and software systems. These biases can result in unfair treatment or outcomes for certain individuals or groups, perpetuating existing disparities and exacerbating societal inequalities. It is essential for companies to scrutinize their programming practices and algorithms to identify and mitigate potential sources of bias, thereby ensuring the integrity and fairness of their systems.

Furthermore, economic disparity can exacerbate existing biases and inequalities within a company, particularly in relation to access to resources, opportunities, and decision-making power. Economic disparities can contribute to unequal outcomes for employees and stakeholders, further widening the gap between privileged and marginalized groups. Addressing economic disparity requires proactive measures to promote equitable access to resources, opportunities, and rewards within the organization.

Overall, systematic bias poses significant challenges to the security and integrity of a company's operations, necessitating a multifaceted approach to address and mitigate its impact. By recognizing and addressing institutional bias, cultural prejudice, discrimination within programming, and economic disparity, companies can foster a more inclusive, equitable, and secure environment for all employees and stakeholders.

Institutional Bias

Institutional bias refers to the systemic tendency for organizations to exhibit preferential treatment or disadvantageous practices toward certain groups or individuals based on inherent biases embedded within their structures, policies, and practices.[540] This form of bias can have significant ramifications for the security dynamics within a company, both negatively and positively.

Negatively, institutional bias can perpetuate inequalities and disparities within the organization, leading to discrimination, exclusion, and mistrust among employees and stakeholders.[541] For example, if hiring practices favor candidates from specific demographic groups or backgrounds, it can result in a lack of diversity and inclusion within the workforce, limiting perspectives and potentially overlooking qualified candidates. Such biases can create friction and resentment among employees, eroding morale and cohesion within the company.

Institutional bias can manifest in decision-making processes, resource allocation, and opportunities for advancement, further exacerbating inequities and hindering the organization's ability to operate securely and effectively.[542] When certain groups feel marginalized or unfairly treated within the company, it can breed resentment and disengagement, ultimately compromising the overall security culture and performance of the organization.

---

[540] Henry, P.J., 2010. Institutional bias. *Handbook of prejudice, stereotyping, and discrimination*, pp.426-440.
[541] *Id.*
[542] Doucouliagos, C., 1995. Institutional bias, risk, and workers' risk aversion. *Journal of Economic Issues*, *29*(4), pp.1097-1118.

However, institutional bias can also have positive implications for security within a company when efforts are made to identify, address, and mitigate biases effectively. By acknowledging and confronting biases within organizational structures and practices, companies can cultivate a more inclusive and equitable environment that fosters trust, collaboration, and innovation among employees.[543] Embracing diversity and promoting equitable opportunities can enhance the company's ability to attract and retain top talent, strengthen its security posture, and drive sustainable growth and success.

Furthermore, by implementing proactive measures to mitigate institutional bias, such as adopting transparent and merit-based hiring practices, promoting diversity and inclusion initiatives, and fostering a culture of accountability and fairness, companies can build resilience against internal and external threats. A commitment to combating institutional bias not only enhances security within the company but also reinforces its reputation as an ethical and responsible organization, thereby bolstering trust and confidence among employees, customers, and stakeholders.

Cultural Prejudice

Cultural prejudice refers to the preconceived notions, stereotypes, or biases that individuals or groups may hold toward people from different cultural backgrounds or identities.[544] These biases can arise from societal norms,

---

[543] Carnes, M., Devine, P.G., Isaac, C., Manwell, L.B., Ford, C.E., Byars-Winston, A., Fine, E. and Sheridan, J., 2012. Promoting institutional change through bias literacy. *Journal of diversity in higher education*, *5*(2), p.63.

[544] Allen, D., Lim, W., Frankel, E., Simons, J., Siddarth, D. and Weyl, G., 2023. Ethics of Decentralized Social Technologies: Lessons from

historical narratives, media portrayals, or personal experiences, and they can significantly impact the security dynamics within a company, both positively and negatively.

Negatively, cultural prejudice can manifest in discriminatory practices, microaggressions, or hostile work environments, where employees from marginalized or minority cultural groups may face barriers to equal opportunities, recognition, or advancement. Such biases can undermine morale, trust, and collaboration among team members, leading to divisions, tensions, and decreased productivity within the company.[545] Additionally, cultural prejudice may contribute to misunderstandings or miscommunications between employees, hindering effective collaboration and decision-making processes, which can compromise the overall security posture of the organization.

However, cultural prejudice can also have positive effects within the security of a company when efforts are made to promote cultural diversity, inclusion, and sensitivity. By embracing diverse perspectives, experiences, and cultural backgrounds, companies can foster innovation, creativity, and resilience within their workforce.[546] Cultivating a culture of respect, empathy, and open-mindedness can enhance communication, collaboration, and problem-solving abilities among employees, ultimately strengthening the company's security culture and capabilities.

Web3, the Fediverse, and Beyond
[545] Moaledj, K., 2023. An Ethnographic Examination of Belonging and Hype in Web3 Communities.
[546] *Id.*

Furthermore, addressing cultural prejudice can improve employee satisfaction, engagement, and retention rates, as individuals from diverse cultural backgrounds feel valued, respected, and included within the organization. By creating an environment where all employees feel empowered to contribute their unique insights and talents, companies can tap into a wealth of diverse perspectives and ideas, driving innovation and competitive advantage in today's global marketplace.[547]

To mitigate the negative effects of cultural prejudice and promote inclusivity, companies can implement diversity and inclusion initiatives, such as cultural competency training, unconscious bias awareness programs, and employee resource groups. By raising awareness of cultural biases and promoting empathy and understanding among employees, companies can create a more inclusive and equitable workplace where everyone can thrive and contribute to the company's success.

Discrimination within the Programing

Discrimination within programming refers to the presence of biases or discriminatory practices embedded within algorithms, software, or technology systems due to the biases of the developers, data used for training, or underlying societal prejudices. These biases can result in unequal treatment or outcomes for individuals based on factors such as race, gender, ethnicity, or other protected characteristics. The problem of discrimination within

[547] Shilina, S., 2023. Decentralized science (DeSci): Web3-mediated future of science. *URL: https://medium. com/paradigm-research/decentralized-science-desci-web3-mediated-future-of-science-2547f9a88c40. Accessed, 12*, p.23.

programming can have significant negative implications for the security of a company across various domains.

Biased algorithms or software can lead to discriminatory outcomes in hiring, promotion, or resource allocation processes within the company. For example, if a recruitment algorithm is trained on historical data that reflects gender or racial biases in hiring decisions, it may perpetuate or exacerbate these biases by favoring certain demographics over others. This can result in a lack of diversity within the company, which not only undermines fairness and equity but also limits access to a diverse talent pool, hindering innovation and competitiveness.

Second, discriminatory programming can erode trust and credibility in the company's products or services, particularly if customers or stakeholders perceive bias or unfair treatment. For instance, if an AI-powered recommendation system exhibits gender or racial bias in its suggestions, it may alienate certain user groups and damage the company's reputation for inclusivity and integrity. Additionally, biased algorithms in cybersecurity systems or fraud detection tools may lead to false positives or negatives, compromising the company's ability to effectively identify and mitigate security threats.

Discrimination within programming can perpetuate societal inequalities and contribute to social tensions or conflicts, which can indirectly impact the security of the company. Biased technology systems that reinforce stereotypes or marginalize certain groups may face public backlash, protests, or legal challenges, resulting in reputational damage, regulatory scrutiny, or financial penalties for the company. Furthermore, discriminatory practices in technology development may create ethical dilemmas for employees, leading to internal discord, employee disengagement, or even whistleblowing actions that expose the company to legal risks or public condemnation.

To address the problem of discrimination within programming and safeguard the security of the company, organizations must prioritize diversity, equity, and inclusion (DEI) efforts across their technology development processes.[548] This includes implementing robust practices for bias detection and mitigation in algorithmic decision-making, such as auditing algorithms for fairness, transparency, and accountability. Additionally, companies should invest in diversity training for developers and data scientists to raise awareness of implicit biases and promote ethical considerations in technology design and deployment. By fostering a culture of diversity and inclusion in technology development, companies can mitigate the risks of discrimination and enhance the security, trust, and resilience of their digital systems.

Economic Disparity

Discrimination based on economic disparity within the corporate structure can significantly impact security within a company, affecting both internal operations and external relationships. Economic inequality often manifests in various forms within organizations, such as unequal access to resources, opportunities, or decision-making power, which can exacerbate disparities and create vulnerabilities that compromise security.[549]

First, economic discrimination may lead to disparities in access to training, professional development, or

---

[548] *Id.*
[549] Smithmyer, *supra.*

advancement opportunities within the company. Employees from economically disadvantaged backgrounds may face barriers to career progression, such as limited access to higher education or networking opportunities, which can hinder their ability to acquire the skills and knowledge necessary for roles involving sensitive or security-related tasks. As a result, the company may have a workforce that lacks the diverse expertise and capabilities needed to effectively address security challenges, leaving gaps in defense mechanisms and increasing susceptibility to threats.

Second, economic disparities can contribute to a culture of mistrust, resentment, or disengagement among employees, which can undermine teamwork, collaboration, and morale within the company. Employees who perceive unfair treatment or unequal opportunities based on economic status may feel undervalued or marginalized, leading to decreased motivation, loyalty, or commitment to organizational goals.[550] This lack of cohesion or solidarity among employees can weaken internal controls, increase the risk of insider threats, or create vulnerabilities that malicious actors may exploit for personal gain or sabotage.

Economic discrimination within the corporate structure can perpetuate power imbalances and inequalities that affect decision-making processes, resource allocation, or compensation practices. Executives or managers from privileged socioeconomic backgrounds may inadvertently prioritize the interests of certain groups or departments over others, leading to inequitable distribution of resources or attention to security concerns. This unequal treatment or neglect of security-related issues based on economic factors

---

[550] *Id.*

can result in overlooked vulnerabilities, inadequate investment in security measures, or ineffective response strategies, leaving the company exposed to potential breaches or disruptions.

To mitigate the security risks associated with discrimination based on economic disparity, companies must prioritize equity, fairness, and inclusivity in their organizational policies, practices, and culture. This includes implementing measures to promote equal access to training, development opportunities, and advancement pathways for all employees, regardless of economic background. Additionally, companies should foster a culture of transparency, accountability, and respect for diversity to address systemic biases, promote trust, and empower employees to contribute to security efforts collaboratively. By addressing economic discrimination and promoting a more inclusive corporate environment, companies can strengthen their security posture, enhance resilience, and foster a culture of shared responsibility and trust among employees.

Conclusion

In conclusion, systematic bias presents multifaceted challenges to security within companies, affecting various aspects of organizational culture, decision-making processes, and resource allocation. Institutional bias, stemming from ingrained organizational norms or structures, can perpetuate inequalities and create vulnerabilities that compromise security. Cultural prejudice, arising from societal attitudes or norms, can foster environments of mistrust or exclusion, hindering collaboration and cooperation among employees and undermining security efforts. Discrimination within programming can introduce biases into algorithms and systems, leading to unintended consequences or vulnerabilities that may be exploited by malicious actors.

Economic disparity within the corporate structure can exacerbate inequalities in access to resources, opportunities, and decision-making power, impacting workforce diversity, morale, and cohesion. Addressing these issues requires a multifaceted approach that encompasses organizational policies, cultural norms, and systemic practices. Companies must strive to cultivate inclusive environments that value diversity, equity, and fairness, fostering cultures of transparency, accountability, and trust. By promoting diversity and inclusion, companies can harness the collective talents and perspectives of their employees to enhance security, resilience, and innovation.

Furthermore, companies should implement measures to mitigate the impact of systematic bias on security, including regular audits and assessments to identify and address biases in organizational policies, practices, and decision-making processes. Training programs and awareness campaigns can raise awareness of bias and discrimination, empowering employees to recognize and challenge discriminatory behaviors or practices. Additionally, companies should foster open communication channels and mechanisms for feedback and grievance redressal, ensuring that employees feel heard, valued, and supported in addressing issues of bias and discrimination.

Overall, addressing systematic bias requires a concerted effort from companies to create inclusive environments, dismantle discriminatory practices, and promote equity and fairness in all aspects of organizational life. By embracing diversity and inclusion as core values and integrating them into their security strategies and practices, companies can enhance their resilience, trustworthiness, and effectiveness in navigating the complex challenges of the digital age.

Chapter 29: Censorship and Management

"There are worse crimes than burning books. One of them is not reading them."

Joseph Brodsky[551]

In corporate messaging, censorship serves as a double-edged sword, wielding both benefits and risks for companies. On one hand, censorship can be viewed as a necessary tool for maintaining a cohesive and consistent message emanating from the company, ensuring alignment with strategic objectives, brand image, and corporate values. By exerting control over the information disseminated to internal and external stakeholders, companies can safeguard their reputation, mitigate risks, and project a unified front to the public. This centralized approach to messaging management can instill confidence among investors, customers, and employees, fostering trust and loyalty in the company's brand.

However, the excessive or indiscriminate application of censorship measures can have unintended consequences, stifling creativity, innovation, and free expression within the organization. When employees feel constrained or inhibited in voicing their ideas, concerns, or dissenting opinions, it can lead to a culture of conformity, complacency, and groupthink, hindering critical thinking, problem-solving, and decision-making processes. Also, an overly restrictive approach to censorship may alienate employees, erode morale, and engender feelings of resentment or disillusionment, ultimately undermining employee engagement, productivity, and retention.

---

[551] (2024). "Censorship Quotes, Joseph Brodsky." Good Reads at https://www.goodreads.com/quotes/tag/censorship.

Furthermore, excessive censorship can create blind spots or gaps in information flow, impeding the timely identification and mitigation of emerging risks, threats, or vulnerabilities. By silencing dissenting voices or suppressing inconvenient truths, companies may inadvertently overlook warning signs or early indicators of potential crises, leaving them ill-prepared to respond effectively when faced with adversity. In an era of rapid technological advancement and digital interconnectedness, where information spreads at lightning speed across various channels and platforms, companies must strike a delicate balance between transparency and discretion in their censorship practices.

Censorship in the digital age extends beyond traditional media channels to encompass social media platforms, online forums, and digital communications channels, presenting unique challenges and opportunities for companies. In an increasingly polarized and volatile online landscape, where misinformation, disinformation, and online harassment abound, companies must navigate carefully to protect their brand reputation while upholding principles of free speech, diversity, and inclusivity. Balancing the need for censorship with respect for individual rights and freedoms requires companies to adopt nuanced approaches that prioritize transparency, accountability, and ethical conduct in their communication strategies.

The Problem of Censorship

Censorship, when wielded without restraint or oversight, can pose significant challenges to the fabric of society and the principles of democracy. At its core, censorship entails the suppression or restriction of information, ideas, or speech deemed objectionable, controversial, or subversive by those in positions of authority. By exerting control over the flow of information, censorship can inhibit the free exchange of ideas, impede intellectual inquiry, and stifle public discourse, thereby hindering the democratic ideals of transparency, accountability, and informed decision-making.

One of the primary concerns associated with censorship is its potential to infringe upon individual rights and freedoms, including the right to freedom of expression and the right to access information.[552] When governments or institutions impose censorship measures to suppress dissenting viewpoints, dissenting voices, or minority perspectives, they undermine the foundational principles of democracy and civil liberties. Censorship can serve as a tool for consolidating power, silencing dissent, and perpetuating inequalities, particularly in authoritarian regimes where freedom of speech is severely curtailed and dissenting voices are systematically suppressed.

Furthermore, censorship can contribute to the erosion of public trust in institutions and undermine the credibility of authoritative sources of information. When information is censored or manipulated to serve vested interests or political agendas, it erodes public confidence in the integrity, impartiality, and reliability of media outlets, government institutions, and other sources of information. This erosion of trust can have far-reaching consequences, including polarization, social division, and a breakdown of social cohesion, as individuals retreat into echo chambers

---

[552] de Vos, M., Ishmaev, G. and Pouwelse, J., 2024. DeScan: Censorship-resistant indexing and search for Web3. *Future Generation Computer Systems*, *152*, pp.257-272.

or alternative information ecosystems that reinforce their existing beliefs and biases.

Censorship can impede progress, innovation, and cultural exchange by stifling creativity, diversity, and intellectual curiosity. When artists, writers, scientists, and scholars are subjected to censorship, it limits their ability to explore new ideas, challenge conventional wisdom, and push the boundaries of knowledge and expression.[553] Inhibiting the free flow of ideas and information deprives society of the diverse perspectives, insights, and innovations needed to address complex challenges and foster social, cultural, and intellectual advancement.

Finally, censorship can have chilling effects on public discourse and self-censorship, as individuals may refrain from expressing their opinions, sharing information, or engaging in open dialogue for fear of reprisal, censorship, or social ostracism. This climate of fear and self-censorship stifles debate, narrows the range of acceptable discourse, and undermines the marketplace of ideas essential for a vibrant and pluralistic society. Ultimately, safeguarding against the perils of censorship requires a commitment to upholding the principles of free speech, transparency, and open inquiry, even in the face of discomfort or disagreement.

The Problem of No Censorship

Within corporations, censorship serves as a means of maintaining a cohesive and consistent message across all levels of the organization. By exerting control over internal communications and external-facing content, companies

[553] Bambacht, J. and Pouwelse, J., 2022. Web3: A decentralized societal infrastructure for identity, trust, money, and data. *arXiv preprint arXiv:2203.00398.*

can ensure that their brand identity, values, and strategic objectives are effectively communicated to employees, stakeholders, and the public. This centralized approach to messaging is essential for fostering a sense of unity, alignment, and clarity within the organization, particularly in fast-paced environments where disparate voices and conflicting narratives can lead to confusion, disengagement, and inefficiency.

Censorship in the corporate context is often driven by the need to protect sensitive information, proprietary knowledge, and intellectual property from unauthorized disclosure or misuse. By implementing strict guidelines and protocols for communication and information sharing, companies can safeguard their competitive advantage, trade secrets, and confidential data from internal leaks or external threats. This proactive approach to information security helps mitigate the risk of corporate espionage, data breaches, and reputational damage, thereby safeguarding the long-term interests and viability of the organization.

Furthermore, censorship plays a crucial role in upholding legal and regulatory compliance within corporations, ensuring that communications adhere to industry standards, best practices, and ethical guidelines. By monitoring and controlling the content disseminated by employees, companies can minimize the risk of inadvertently violating laws, regulations, or contractual obligations related to advertising, marketing, data privacy, and consumer protection.[554] Compliance with legal and regulatory requirements not only protects the company from potential litigation or regulatory penalties but also fosters trust and

---

[554] *Id.*

confidence among customers, investors, and other stakeholders.

Moreover, censorship within corporations can help mitigate the spread of misinformation, rumors, or misleading information that could undermine employee morale, productivity, and organizational cohesion. By vetting and verifying the accuracy of communications, companies can prevent the dissemination of false or misleading information that may sow confusion, erode trust, or damage the company's reputation. This proactive approach to managing internal communications promotes transparency, credibility, and accountability, reinforcing the company's commitment to integrity and ethical conduct.

Additionally, censorship serves as a tool for promoting brand consistency and reputation management, ensuring that all communications align with the company's values, mission, and brand identity. By enforcing standards for tone, messaging, and imagery, companies can maintain a strong and unified brand presence across various channels and touchpoints, enhancing brand recognition, loyalty, and trust among customers and stakeholders. Consistent messaging also helps mitigate the risk of misinterpretation, misrepresentation, or reputational harm that could arise from inconsistent or contradictory communications.

Furthermore, censorship within corporations can facilitate effective crisis communication and risk management, enabling organizations to respond swiftly and decisively to emerging threats, controversies, or crises. By controlling the flow of information and coordinating messaging efforts during times of crisis, companies can mitigate the impact of negative publicity, address stakeholder concerns, and protect their reputation and brand equity. Timely and transparent communication, guided by a carefully crafted crisis communication plan, can help restore public trust, minimize reputational damage, and preserve shareholder value in the face of adversity.

Censorship within corporations plays a multifaceted role in shaping internal and external communications, safeguarding sensitive information, ensuring legal compliance, fostering brand consistency, and managing risks and crises effectively. While censorship may raise concerns about transparency, accountability, and freedom of expression, when implemented judiciously and ethically, it can serve as a valuable tool for promoting unity, clarity, and trust within organizations, ultimately contributing to their long-term success and sustainability.

Balancing Between Too Much and Too Little Censorship

Balancing censorship within a company requires careful consideration of various factors, including the organization's culture, values, industry norms, and legal obligations. On one hand, excessive censorship can stifle creativity, innovation, and employee engagement, leading to a culture of fear, mistrust, and disengagement. When employees feel constrained or censored in expressing their ideas, opinions, or concerns, it can hinder collaboration, problem-solving, and decision-making processes, ultimately impeding the company's ability to adapt, innovate, and thrive in a dynamic business environment.

Conversely, insufficient censorship or lax control over communications can pose significant risks to the company's reputation, brand integrity, and legal compliance.[555]

---

[555] Fecher, B., Kuper, F., Fähnrich, B., Schmid-Petri, H., Schildhauer, T., Weingart, P. and Wormer, H., 2023. Balancing interests between freedom and censorship: Organizational strategies for quality assurance

Without clear guidelines and oversight mechanisms in place, employees may inadvertently disclose sensitive information, make unauthorized claims, or engage in behavior that could harm the company's interests or expose it to legal liabilities. Moreover, unchecked communications can lead to inconsistencies, contradictions, or misrepresentations that undermine the company's credibility, trustworthiness, and stakeholder relationships, potentially eroding shareholder value and market standing.

To strike the right balance between too much and too little censorship, companies must adopt a nuanced and context-sensitive approach that recognizes the importance of transparency, openness, and accountability while also prioritizing confidentiality, security, and compliance. This entails establishing clear policies, procedures, and guidelines for communication and information sharing, which are informed by legal requirements, industry standards, and ethical principles. Moreover, fostering a culture of trust, respect, and inclusivity where employees feel empowered to voice their opinions, raise concerns, and contribute to meaningful dialogue can help mitigate the need for excessive censorship while promoting a healthy and productive work environment.[556] By promoting a culture of responsible communication and providing employees with the tools, training, and support they need to navigate complex ethical and legal dilemmas, companies can navigate the delicate balance between censorship and freedom of expression, ensuring that their communications remain both transparent and secure.

---

in science communication. *Science and Public Policy*, *50*(1), pp.1-14.
[556] *Id.*

Conclusion

The management of censorship within corporations is a complex endeavor that requires careful consideration of various factors, including the need to maintain a unified message while fostering an environment conducive to creativity, innovation, and employee engagement. While censorship can serve important purposes in aligning corporate communications with strategic objectives, brand values, and legal requirements, it must be balanced with the principles of transparency, openness, and respect for diverse perspectives. Striking this balance requires clear policies, guidelines, and oversight mechanisms that promote responsible communication while safeguarding against risks such as misinformation, reputational damage, and legal liabilities.

The management of censorship should not be seen as a one-size-fits-all approach but rather as a dynamic and context-sensitive process that evolves in response to changing circumstances, stakeholder expectations, and industry trends. Companies must continuously reassess their censorship practices to ensure they remain relevant, effective, and aligned with their broader goals and values. This may involve periodic reviews, audits, and updates to policies, as well as ongoing communication and training initiatives to reinforce desired behaviors and mitigate potential risks.

Furthermore, while censorship can play a crucial role in maintaining brand consistency, protecting sensitive information, and complying with regulatory requirements, it should never be used as a means of suppressing dissent, silencing criticism, or undermining employee autonomy. Companies must strive to cultivate a culture of openness, trust, and accountability where employees feel empowered to express their views, challenge assumptions, and contribute to meaningful dialogue without fear of reprisal or censorship.

Ultimately, effective censorship management requires a delicate balance between the need for control and the imperative of fostering a culture of transparency, authenticity, and ethical communication.[557] By embracing this balance and adopting a proactive and inclusive approach to censorship management, companies can navigate the complexities of corporate communications while upholding their commitments to integrity, responsibility, and respect for all stakeholders involved.

[557] *Id.*

Conclusion Part VI

In conclusion, Part VI has delved into the multifaceted challenges posed by algorithmic bias, systematic bias, and censorship within the realm of corporate security. These chapters have underscored the importance of recognizing and addressing the various forms of bias and censorship that can undermine the integrity, trustworthiness, and effectiveness of security measures within organizations operating in the digital landscape. By understanding these issues and implementing appropriate strategies to mitigate their impact, companies can enhance their security posture and foster a culture of inclusivity, transparency, and ethical responsibility.

Chapter 27 explored the phenomenon of algorithmic bias and its implications for security, highlighting how discrimination, privacy erosion, and vulnerability to attacks can result from biased algorithms. By acknowledging the existence of algorithmic bias and taking proactive steps to mitigate its effects, companies can minimize the risk of unintended consequences and uphold principles of fairness, equity, and accountability in their operations.

Chapter 28 examined the concept of systematic bias, emphasizing the importance of addressing institutional bias, cultural prejudice, discrimination in programming, and economic disparity within corporate environments. By promoting diversity, equity, and inclusion initiatives and implementing measures to combat bias and discrimination, organizations can create a more inclusive and resilient security culture that reflects the values and aspirations of all stakeholders.

Chapter 29 focused on the management of censorship within corporate communications, highlighting the delicate balance between the need for a unified message and the imperative of preserving transparency, authenticity, and freedom of expression. By adopting a nuanced approach to censorship management and fostering an environment that encourages open dialogue, constructive feedback, and diverse perspectives, companies can navigate the complexities of corporate communications while upholding their commitments to integrity and ethical conduct.

Together, these chapters underscore the interconnected nature of security, ethics, and organizational culture, emphasizing the importance of adopting a holistic approach to security management that addresses both technical and socio-cultural dimensions. By integrating principles of fairness, accountability, and respect for human rights into their security practices, companies can build trust, legitimacy, and loyalty among employees, customers, and other stakeholders, thereby enhancing their resilience and competitiveness in an increasingly interconnected and dynamic digital landscape.

As we move forward, it is essential for organizations to remain vigilant and adaptive in their approach to security, continuously evaluating and refining their strategies in response to emerging threats, technological advancements, and evolving regulatory landscapes. By fostering a culture of continuous improvement, innovation, and collaboration, companies can effectively navigate the complexities of the digital age while safeguarding the trust and confidence of their stakeholders.

Part VII: Responsible Use of AI and Machine Learning

Introduction Part VII

Part VII of our comprehensive exploration into security in the digital environment. In this installment, we delve into the dynamic intersection of artificial intelligence (AI) and machine learning within the realm of cybersecurity, focusing on their pivotal role in investment and development monitoring for web3 companies. As technology continues to evolve at an unprecedented pace, so too do the complexities and challenges of securing digital assets and navigating the digital landscape. Part VII serves as a beacon guiding us through the intricate interplay between technological innovation and cybersecurity strategies, shedding light on the transformative potential of AI and machine learning in safeguarding digital ecosystems.

Chapter 30 serves as our foundational cornerstone, laying the groundwork for our journey by providing a comprehensive overview of cybersecurity fundamentals. From understanding the anatomy of cyber threats to implementing robust defense mechanisms, this chapter equips readers with the essential knowledge and insights needed to navigate the ever-changing cybersecurity landscape. We explore the importance of proactive risk management strategies and the imperative for web3 companies to prioritize cybersecurity in their operations.

Building upon this foundational understanding, Chapter 31 delves into the realm of data privacy and compliance in the digital age. In an era marked by unprecedented data proliferation and regulatory scrutiny, organizations face mounting pressure to uphold stringent data privacy standards and navigate complex regulatory frameworks. This chapter explores the evolving regulatory landscape, ethical considerations, and best practices for ensuring compliance and fostering trust among stakeholders.

Chapter 32 marks a pivotal turning point in our exploration, as we delve into the convergence of AI and cybersecurity. As AI-driven technologies continue to revolutionize cybersecurity practices, organizations are increasingly turning to machine learning algorithms for threat detection, incident response, and vulnerability management. We examine the transformative potential of AI in augmenting cybersecurity defenses, enhancing threat intelligence capabilities, and enabling proactive defense strategies.

Finally, Chapter 33 propels us into the forefront of innovation, as we explore the intersection of AI and machine learning with investment and development monitoring. In today's digital landscape, web3 companies are leveraging investment bots, automated trading algorithms, and AI-driven analytics to optimize financial management practices and drive innovation. This chapter delves into the intricacies of investment strategies, portfolio management, and the transformative potential of AI in shaping the future of financial markets and development initiatives.

Part VII offers a comprehensive journey through the dynamic interplay between AI, machine learning, and cybersecurity within the realm of investment and development monitoring for web3 companies. As we embark on this journey, we invite readers to join us in exploring the transformative potential of these technologies and their implications for securing digital ecosystems in an era defined by rapid technological advancement and digital innovation.

Chapter 30: Transparency of AI Use

"The development of full artificial intelligence could spell the end of the human race....It would take off on its own, and re-design itself at an ever increasing rate. Humans, who are limited by slow biological evolution, couldn't compete, and would be superseded."

Stephen Hawking[558]

The integration of Artificial Intelligence (AI) into various facets of technology has become ubiquitous, impacting both digital and physical security domains. The transparency of AI usage emerges as a critical concern, affecting trust, accountability, and overall security measures. Ensuring transparency in AI utilization is paramount, given its pervasive influence on decision-making processes and operational functionalities across diverse sectors. This chapter delves into the significance of transparency in AI use, examining its profound implications for digital and physical security realms.

Trust and accountability are foundational elements in the deployment of AI systems, especially in security-critical contexts. Transparency plays a pivotal role in fostering trust among stakeholders, including users, regulators, and the general public. By providing clear insights into the functioning and decision-making processes of AI algorithms, organizations can enhance accountability and establish trust in their security mechanisms. Moreover, transparency facilitates greater scrutiny and oversight, enabling stakeholders to assess the reliability and fairness of AI-driven security measures effectively.

---

558 (2024). "Quotes on AI, Stephen Hawking." Bernard Marr & Co at https://bernardmarr.com/28-best-quotes-about-artificial-intelligence/.

AI bias and discrimination pose significant challenges to security, potentially exacerbating existing vulnerabilities and disparities. Transparent AI practices are instrumental in mitigating bias and discrimination, as they allow for the identification and rectification of algorithmic shortcomings. By shedding light on the underlying data sources, training methodologies, and decision-making criteria, transparency empowers organizations to address biases proactively and promote equity in security outcomes. Furthermore, transparency fosters awareness of potential biases among developers, encouraging the adoption of inclusive and unbiased AI models.

Safe and secure development of AI systems is essential to safeguarding digital and physical security infrastructures from potential threats and vulnerabilities. Transparency serves as a cornerstone in the development process, facilitating peer review, collaboration, and knowledge sharing. Openness in AI development practices enables researchers and practitioners to identify security loopholes, validate algorithmic robustness, and implement effective countermeasures against emerging threats. By promoting transparency, organizations can fortify their AI systems against adversarial attacks, data breaches, and other security risks, thereby enhancing overall system resilience.

Promotion of responsible innovation is imperative to harnessing the full potential of AI while mitigating associated risks and uncertainties. Transparency in AI usage fosters responsible innovation by fostering informed decision-making, ethical considerations, and stakeholder engagement. By openly disclosing AI capabilities, limitations, and intended use cases, organizations can facilitate constructive dialogue and collaboration among diverse stakeholders, including policymakers, industry experts, and civil society representatives. Transparent AI practices engender a culture of responsible innovation, wherein ethical, legal, and societal implications are systematically addressed to ensure positive societal impact.

Transparency in AI utilization is indispensable for ensuring digital and physical security in today's interconnected world. By prioritizing trust, accountability, bias mitigation, secure development, and responsible innovation, organizations can navigate the complexities of AI adoption while safeguarding critical security infrastructures. This chapter delves into the multifaceted nature of transparency in AI usage, highlighting its pivotal role in shaping the future of security and technology. Through collaborative efforts and adherence to transparent AI practices, stakeholders can collectively work towards building resilient and trustworthy security ecosystems.

Building Trust and Accountability

Trust and accountability are paramount in the realm of AI, particularly concerning the security of the company's operations and data. Firstly, trust in AI systems is essential for fostering confidence among stakeholders, including employees, customers, and business partners. When individuals trust that AI technologies are reliable, accurate, and transparent in their decision-making processes, they are more likely to embrace and utilize these systems for security-related tasks, such as threat detection, access control, and incident response. By instilling trust in AI, companies can enhance the effectiveness and adoption of security measures, thereby fortifying their defenses against potential risks and vulnerabilities.

Accountability in AI is crucial for ensuring that decisions made by automated systems align with corporate values, legal requirements, and ethical standards.[559] Companies must establish clear lines of responsibility and oversight to hold AI systems accountable for their actions and outcomes, particularly in security-critical domains where errors or biases could have significant repercussions. By implementing robust governance frameworks and mechanisms for monitoring and auditing AI algorithms, organizations can mitigate the risk of unintended consequences, such as data breaches, privacy violations, or discriminatory practices, which could undermine trust and tarnish the company's reputation.

Furthermore, trust and accountability in AI contribute to the development of a culture of responsibility and integrity within the organization, where employees are empowered to use AI tools responsibly and ethically to support security objectives. When individuals understand the principles and processes behind AI decision-making and are confident in the reliability and fairness of these systems, they are more likely to uphold ethical standards and comply with security protocols in their daily activities.[560] By promoting a culture of trust and accountability, companies can cultivate a workforce that is invested in safeguarding the company's assets, reputation, and stakeholders' interests, thereby strengthening the overall security posture.

Ultimately, trust and accountability in AI serve as foundational pillars for building resilience and sustainability in the company's security practices. By prioritizing transparency, integrity, and ethical behavior in

---

[559] Jacovi, A., Marasović, A., Miller, T. and Goldberg, Y., 2021, March. Formalizing trust in artificial intelligence: Prerequisites, causes and goals of human trust in AI. In *Proceedings of the 2021 ACM conference on fairness, accountability, and transparency* (pp. 624-635).
[560] Bedué, P. and Fritzsche, A., 2022. Can we trust AI? An empirical investigation of trust requirements and guide to successful AI adoption. *Journal of Enterprise Information Management*, *35*(2), pp.530-549.

the design, deployment, and use of AI technologies, organizations can instill confidence in stakeholders, mitigate risks associated with automated decision-making, and uphold their commitment to protecting sensitive data and assets. Through proactive efforts to foster trust and accountability, companies can harness the potential of AI to enhance security outcomes while minimizing the potential for harm or misuse, thereby advancing their strategic objectives and ensuring long-term success.

Mitigating AI Bias and Discrimination

Mitigating AI bias and discrimination is imperative for ensuring digital security within web3 companies. Firstly, organizations can implement rigorous data collection and preprocessing techniques to identify and address biases present in training datasets. By carefully curating and diversifying training data sources, companies can reduce the risk of perpetuating or amplifying existing biases in AI algorithms, thus promoting fairness and inclusivity in decision-making processes.[561] Additionally, employing techniques such as data augmentation, bias quantification, and algorithmic debiasing can help mitigate biases by promoting a more balanced representation of different demographic groups and minimizing the impact of biased features on model predictions.

Furthermore, companies can prioritize transparency and interpretability in AI systems to enhance accountability and mitigate the potential for discriminatory outcomes. By providing stakeholders with insights into the underlying mechanisms and decision-making processes of AI

---

[561] Ferrer, X., Van Nuenen, T., Such, J.M., Coté, M. and Criado, N., 2021. Bias and discrimination in AI: a cross-disciplinary perspective. *IEEE Technology and Society Magazine*, *40*(2), pp.72-80.

algorithms, organizations can increase awareness and understanding of how biases may manifest in algorithmic outputs. This transparency enables stakeholders to identify and challenge discriminatory practices, fostering a culture of accountability and oversight that holds AI systems accountable for their actions. Additionally, promoting diversity and inclusivity in AI development teams can help mitigate bias by bringing diverse perspectives and experiences to the design and evaluation of AI systems, thereby reducing the likelihood of overlooking or perpetuating biases.

Companies can establish clear policies and guidelines for ethical AI use, incorporating principles of fairness, transparency, and accountability into the design, deployment, and evaluation of AI systems. By adopting frameworks such as the AI Ethics Guidelines developed by leading organizations and regulatory bodies, companies can outline best practices for mitigating bias and discrimination in AI, including measures for auditing and monitoring AI systems for discriminatory behavior.[562] These policies should emphasize the importance of fairness, equity, and respect for human rights in AI decision-making, while also providing mechanisms for remediation and redress in cases of discrimination or bias-related harm.

Last, ongoing monitoring, evaluation, and validation of AI systems are essential for identifying and addressing biases and discriminatory practices in real-world contexts. By regularly assessing the performance and impact of AI algorithms across different demographic groups and use cases, companies can proactively detect and remediate biases before they result in adverse outcomes or harm. This

---

[562] Varona, D. and Suárez, J.L., 2022. Discrimination, bias, fairness, and trustworthy AI. *Applied Sciences*, *12*(12), p.5826.

continuous monitoring process should involve collaboration between data scientists, domain experts, and affected stakeholders to ensure that AI systems operate ethically and equitably in diverse and dynamic environments. Through these proactive measures, web3 companies can mitigate AI bias and discrimination, fostering a more inclusive and secure digital ecosystem for all users.

Enabling Safe and Secure Development

Artificial intelligence (AI) plays a crucial role in enhancing the safe and secure development of high-sensitivity materials within web3 companies. Firstly, AI-powered anomaly detection systems can continuously monitor and analyze data streams to identify any irregularities or potential security threats in real-time. By leveraging machine learning algorithms, these systems can detect deviations from normal patterns and alert security personnel to investigate and mitigate potential risks promptly.[563] This proactive approach enables companies to detect and respond to security incidents more effectively, thereby minimizing the likelihood of unauthorized access or data breaches.

Second, AI-driven encryption techniques can bolster the security of sensitive materials by providing robust cryptographic protection against unauthorized access or tampering. Advanced encryption algorithms, coupled with AI-based key management systems, can strengthen data protection measures and mitigate the risk of data exposure or leakage. Additionally, AI-enabled encryption solutions

[563] Brundage, M., Avin, S., Wang, J., Belfield, H., Krueger, G., Hadfield, G., Khlaaf, H., Yang, J., Toner, H., Fong, R. and Maharaj, T., 2020. Toward trustworthy AI development: mechanisms for supporting verifiable claims. *arXiv preprint arXiv:2004.07213.*

can dynamically adapt encryption parameters based on contextual factors such as user behavior, network conditions, and threat levels, ensuring adaptive and resilient security defenses against evolving cyber threats.

Furthermore, AI can facilitate the development of sophisticated access control mechanisms tailored to the specific security requirements of high-sensitivity materials. By analyzing historical access patterns and user behavior data, AI-powered access control systems can dynamically adjust access permissions and authentication requirements to enforce least privilege principles and prevent unauthorized access. Additionally, AI-driven biometric authentication technologies, such as facial recognition and voice authentication, can provide enhanced security measures while ensuring user convenience and usability.

AI-based risk assessment and threat modeling tools can assist web3 companies in identifying and prioritizing potential security risks associated with high-sensitivity materials. By leveraging machine learning algorithms to analyze vast amounts of data and identify patterns indicative of security vulnerabilities or attack vectors, these tools can help organizations proactively mitigate risks and strengthen their security posture.[564] Additionally, AI-driven predictive analytics can enable companies to anticipate and mitigate emerging threats before they escalate into full-blown security incidents, thereby reducing the likelihood of data breaches or operational disruptions.

Additionally, AI-powered incident response and recovery systems can enhance the resilience of web3 companies against security breaches or cyber attacks targeting high-

---

[564] Barmer, H., Dzombak, R., Gaston, M., Heim, E., Palat, V., Redner, F., Smith, T. and VanHoudnos, N., 2021. Robust and Secure AI.

sensitivity materials. By automating incident detection, analysis, and response processes, these systems can enable rapid incident containment and mitigation, minimizing the impact of security incidents on business operations and data integrity. Furthermore, AI-driven forensic analysis tools can facilitate post-incident investigations by uncovering root causes and identifying perpetrators, enabling companies to take corrective actions and prevent future security incidents effectively.

Overall, the strategic integration of AI technologies into the development and management of high-sensitivity materials can significantly enhance the safety and security of web3 companies' digital assets. By leveraging AI-driven solutions for anomaly detection, encryption, access control, risk assessment, and incident response, organizations can establish robust security frameworks that protect sensitive materials from unauthorized access, data breaches, and cyber threats. As AI continues to evolve and advance, its role in ensuring the safe and secure development of high-sensitivity materials will become increasingly indispensable in the web3 era.

Promoting Responsible Innovation

Artificial intelligence (AI) holds immense potential for reviewing policies and plans to ensure safe innovation within web3 companies, particularly in safeguarding the data of the company and its customers. Firstly, AI-powered policy analysis tools can analyze existing policies and procedures to identify potential gaps or areas of improvement related to data security and privacy. By leveraging natural language processing (NLP) algorithms, these tools can parse through complex policy documents and highlight key areas that may require revision or reinforcement to align with emerging regulatory requirements or industry best practices.

Furthermore, AI-driven risk assessment frameworks can evaluate the potential impact of proposed innovations on data security and privacy, enabling companies to make informed decisions about their implementation. By analyzing various risk factors, such as data sensitivity, regulatory compliance, and threat landscape, these AI-powered frameworks can provide actionable insights into the potential risks and vulnerabilities associated with new initiatives or technologies. This proactive approach allows companies to address security concerns upfront and implement appropriate safeguards to protect sensitive data and mitigate potential risks effectively.

AI-based predictive modeling techniques can forecast the potential impact of innovation on data security and privacy, enabling companies to anticipate and mitigate emerging threats or vulnerabilities proactively. By analyzing historical data and trend patterns, these models can identify potential risk factors and predict future outcomes, allowing organizations to implement preemptive measures to safeguard their data assets and maintain compliance with regulatory requirements. Additionally, AI-driven scenario planning tools can simulate various scenarios and assess their potential implications on data security and privacy, enabling companies to develop robust contingency plans and response strategies.

Additionally, AI-powered compliance monitoring systems can continuously monitor and audit data handling practices to ensure adherence to regulatory requirements and internal policies. By automating the detection of non-compliant behavior and anomalous activities, these systems can identify potential data security breaches or policy violations in real-time, enabling companies to take prompt corrective actions and prevent potential risks or liabilities. Furthermore, AI-driven compliance frameworks can provide proactive alerts and recommendations to stakeholders, facilitating ongoing compliance efforts and fostering a culture of accountability and transparency within the organization.

Furthermore, AI-driven customer data protection solutions can enhance the security and privacy of customer data by detecting and preventing unauthorized access or misuse. By leveraging machine learning algorithms to analyze user behavior and transaction patterns, these solutions can identify suspicious activities or potential security threats, enabling companies to intervene proactively and protect customer data from unauthorized access or exploitation. Additionally, AI-powered data anonymization techniques can help companies anonymize sensitive customer information to mitigate the risk of data breaches while preserving data utility for analysis and innovation purposes.

AI technologies offer powerful capabilities for reviewing policies and plans to ensure safe innovation within web3 companies, particularly in safeguarding the data of the company and its customers. By leveraging AI-driven policy analysis, risk assessment, predictive modeling, compliance monitoring, and customer data protection solutions, organizations can proactively identify and mitigate potential risks to data security and privacy, fostering a culture of responsible innovation and trust among stakeholders. As AI continues to evolve and advance, its role in ensuring safe and secure innovation within web3 companies will become increasingly indispensable in the digital age.

Conclusion

The integration of AI into the development and management of web3 companies holds significant promise for enhancing security, privacy, and innovation. By leveraging AI-powered tools and techniques, organizations can proactively identify and address potential risks and vulnerabilities in their policies, plans, and data handling practices. From policy analysis and risk assessment to compliance monitoring and customer data protection, AI enables companies to strengthen their security posture and ensure the responsible handling of sensitive information.

However, it is essential to recognize that the deployment of AI in web3 companies is not without challenges and potential pitfalls. As AI systems become increasingly sophisticated and autonomous, they may introduce new complexities and risks, including algorithmic bias, discrimination, and privacy violations. Therefore, it is crucial for companies to implement robust governance frameworks and ethical guidelines to guide the development and deployment of AI technologies responsibly.

The success of AI implementation in web3 companies depends on fostering a culture of collaboration, transparency, and accountability across all levels of the organization. By involving stakeholders in the decision-making process and promoting open communication channels, companies can ensure that AI technologies are deployed ethically and in alignment with corporate values and regulatory requirements.

Furthermore, continuous monitoring and evaluation of AI systems are essential to identify and address emerging threats or vulnerabilities promptly. Companies must invest in ongoing training and education programs to equip employees with the knowledge and skills needed to understand and mitigate the risks associated with AI technologies effectively.

Overall, AI represents a powerful tool for driving innovation and enhancing security within web3 companies. By embracing AI-driven solutions and adopting a proactive approach to risk management, organizations can navigate the complexities of the digital landscape while safeguarding their data and maintaining the trust and confidence of their stakeholders.

Chapter 31: Bias Mitigation

"What all of us need to do is make sure we are using AI in a way that is for the benefit of humanity, not to the detriment of humanity."

Tim Cook[565]

Addressing bias has emerged in artificial intelligence (AI) and machine learning (ML), as a critical priority for organizations seeking to enhance security and fairness in their systems. AI bias refers to the systematic errors or prejudices that can manifest in algorithms, leading to discriminatory outcomes or skewed results. As web3 companies increasingly rely on AI-driven technologies to support their operations and decision-making processes, the need for effective bias mitigation measures becomes paramount to ensure equitable treatment and safeguard against potential risks.

Native bias mitigation involves identifying and addressing inherent biases present within AI models and algorithms. These biases may stem from the data used to train the algorithms, reflecting historical patterns of discrimination or underrepresentation. By employing techniques such as data preprocessing, feature selection, and algorithmic adjustments, organizations can strive to mitigate the influence of native biases and promote more equitable outcomes. Additionally, ongoing monitoring and evaluation of AI systems are essential to detect and rectify bias-related issues as they arise.

Conditional bias mitigation focuses on addressing biases that may arise under specific conditions or contexts within

---

565 (2024). "AI Quotes, Tim Cook" Brainy Quote at https://www.brainyquote.com/topics/ai-quotes.

AI systems. These biases can result from the interaction between the algorithm and external factors, such as user demographics, environmental variables, or task-specific requirements. To mitigate conditional biases, companies must carefully design their AI models to account for diverse scenarios and mitigate potential sources of discrimination. This may involve implementing fairness-aware algorithms, incorporating bias detection mechanisms, and fine-tuning model parameters to optimize performance across different conditions.

Removing bias-based factors entails identifying and eliminating sources of bias from the data, algorithms, and decision-making processes underlying AI systems. This approach requires a comprehensive analysis of the entire AI pipeline, from data collection and preprocessing to model training and deployment. By implementing rigorous data validation and quality assurance procedures, organizations can identify and mitigate bias at each stage of the AI lifecycle. Furthermore, fostering diversity and inclusivity in AI development teams can help mitigate the risk of unintentional bias and promote the development of fairer and more representative systems.

AI bias mitigation plays a crucial role in enhancing security and fairness within web3 companies' AI-driven systems. By proactively addressing native biases, mitigating conditional biases, and removing bias-based factors, organizations can promote equitable outcomes and reduce the risk of discriminatory practices. However, achieving effective bias mitigation requires a multifaceted approach that encompasses data governance, algorithmic transparency, and stakeholder engagement. As web3 companies continue to leverage AI technologies to drive innovation and growth, prioritizing bias mitigation efforts will be essential to building trust, ensuring compliance, and upholding ethical standards in the digital age.

Native Bias Mitigation

Addressing native bias in AI programs is imperative for companies striving to maintain security while upholding corporate goals and values.[566] Firstly, native bias can inadvertently perpetuate discriminatory practices, leading to unequal treatment or access to security measures based on demographic or other protected characteristics. This misalignment with corporate goals undermines efforts to foster inclusivity and fairness, potentially resulting in legal ramifications and reputational damage. Therefore, companies must proactively identify and mitigate bias in their AI programs to uphold their commitment to equity and non-discrimination in security practices.

Native bias in AI programs can compromise the effectiveness of security measures, rendering them less reliable and robust in safeguarding against emerging threats. If AI algorithms exhibit bias towards certain groups or fail to account for diverse scenarios, security vulnerabilities may go undetected or inadequately addressed, exposing the company to heightened risks of data breaches, cyberattacks, or other security breaches.[567] By diligently scrutinizing AI systems for native bias and ensuring alignment with corporate security objectives,

---

[566] Schwartz, R., Schwartz, R., Vassilev, A., Greene, K., Perine, L., Burt, A. and Hall, P., 2022. *Towards a standard for identifying and managing bias in artificial intelligence* (Vol. 3, p. 00). US Department of Commerce, National Institute of Standards and Technology.
[567] Zhiltsova, A., Caton, S. and Mulwa, C., 2019. Mitigation of unintended biases against non-native English texts in sentiment analysis.

organizations can enhance the efficacy and resilience of their security infrastructure.

Furthermore, failing to address native bias in AI programs can erode trust and confidence in security measures among stakeholders, including employees, customers, and regulatory authorities. If individuals perceive security practices as unfair, discriminatory, or biased, they may become reluctant to engage with security protocols or disclose sensitive information, undermining the company's ability to protect its assets and data effectively.[568] To maintain trust and credibility, companies must prioritize the elimination of bias in AI programs and cultivate a security culture that values transparency, integrity, and equity in all aspects of operations.

Additionally, unchecked native bias in AI programs can lead to unintended consequences and ethical dilemmas, potentially resulting in unintended harm or negative impact on individuals or communities. By overlooking or disregarding biases in AI algorithms, companies risk perpetuating systemic inequalities and exacerbating social disparities, which can have far-reaching implications for both internal and external stakeholders. Therefore, companies must adopt a proactive approach to identify and mitigate bias in their AI programs, guided by ethical considerations and a commitment to social responsibility in security practices.

Addressing native bias in AI programs is essential for companies seeking to uphold their corporate goals related to security while fostering inclusivity, reliability, and trustworthiness in their operations. By prioritizing the identification, evaluation, and mitigation of bias in AI

---

[568] Schwartz, *supra*.

algorithms, organizations can enhance the effectiveness, fairness, and integrity of their security measures, thereby bolstering their resilience against evolving threats and earning the trust and confidence of stakeholders. Through continuous vigilance and ethical oversight, companies can navigate the complexities of AI-driven security solutions while promoting equity, transparency, and accountability in their business practices.

Conditional Bias Mitigation

Conditional bias mitigation is crucial for protecting security in a web3 company as it ensures that AI systems deliver fair and equitable outcomes across diverse contexts and scenarios. In the dynamic and evolving landscape of web3, where companies interact with a wide range of users, stakeholders, and environments, the potential for bias to manifest under specific conditions is significant. By addressing conditional biases, organizations can mitigate the risk of discriminatory practices and uphold the principles of fairness, transparency, and accountability in their AI-driven solutions.[569]

One key reason why conditional bias mitigation is essential for web3 companies' security is its role in promoting trust and confidence among users and stakeholders. When AI systems exhibit biases or unfair treatment in certain contexts, it can erode trust in the company's products, services, and brand reputation. By proactively identifying and mitigating conditional biases, organizations demonstrate their commitment to providing inclusive and unbiased experiences for all users, thereby enhancing trust

[569] Fu, R., Huang, Y. and Singh, P.V., 2020. Ai and algorithmic bias: Source, detection, mitigation and implications. *Detection, Mitigation and Implications (July 26, 2020).*

and fostering positive relationships with customers, partners, and the broader community.

Conditional bias mitigation is critical for ensuring compliance with regulatory requirements and industry standards related to data privacy, anti-discrimination laws, and ethical AI practices. In the highly regulated environment of web3, where companies handle sensitive data and operate in diverse geographical regions, adherence to legal and ethical guidelines is paramount.[570] By implementing robust conditional bias mitigation measures, organizations can demonstrate their commitment to compliance and minimize the risk of legal liabilities, fines, and reputational damage associated with biased AI systems.

Furthermore, conditional bias mitigation contributes to enhancing the effectiveness and reliability of AI-driven security solutions deployed by web3 companies. In security-critical domains such as fraud detection, threat analysis, and identity verification, the accuracy and fairness of AI algorithms are of utmost importance. By addressing conditional biases that may arise in specific use cases or environments, organizations can improve the performance and efficacy of their AI systems, thereby strengthening their overall security posture and resilience against emerging threats and vulnerabilities.

Additionally, conditional bias mitigation fosters innovation and creativity within web3 companies by promoting diversity of perspectives and inclusive decision-making processes. By acknowledging and addressing biases that may affect certain groups or communities, organizations

---

[570] Zhang, Y., Li, B., Ling, Z. and Zhou, F., 2024, March. Mitigating label bias in machine learning: Fairness through confident learning. In *Proceedings of the AAAI Conference on Artificial Intelligence* (Vol. 38, No. 15, pp. 16917-16925).

can unlock new opportunities for collaboration, co-creation, and value generation. This inclusive approach not only drives innovation but also helps web3 companies build stronger and more resilient ecosystems that are better equipped to address complex security challenges and adapt to evolving threats in the digital landscape.

Removing Biased Bias Mitigation Factors

Removing bias factors created by the bias against bias for the creators of AI is crucial for ensuring the integrity and effectiveness of AI systems in web3 companies. While efforts to mitigate bias are essential, it is equally important to recognize that certain factors are based on objective data and facts rather than discriminatory intent. In many cases, biases against bias may inadvertently lead to the suppression or dismissal of legitimate factors that are relevant to decision-making processes.

One approach to addressing this challenge is to adopt a nuanced and context-aware approach to bias mitigation that acknowledges the complexity of the data landscape and the inherent trade-offs involved in algorithmic decision-making. Rather than categorically eliminating certain factors or attributes based on perceived biases, creators of AI systems should engage in thorough analysis and validation processes to distinguish between genuine discriminatory patterns and valid indicators of risk, relevance, or predictive value.

Moreover, removing bias factors requires a careful balancing act between the imperatives of fairness, accuracy, and utility in AI systems. While it is essential to minimize the impact of discriminatory biases on decision outcomes, it is equally important to ensure that AI algorithms remain effective, efficient, and aligned with the objectives of the organization.[571] This requires a nuanced understanding of the underlying data dynamics and the potential implications of bias mitigation strategies on the overall performance and reliability of AI systems.

Furthermore, fostering transparency and accountability in the development and deployment of AI technologies can help mitigate the risk of bias factors stemming from biases against bias. By adopting transparent practices and providing stakeholders with insights into the data sources, algorithms, and decision-making processes underlying AI systems, creators can enhance trust, promote informed decision-making, and facilitate meaningful dialogue around the ethical and social implications of AI-driven solutions. Additionally, establishing robust mechanisms for ongoing monitoring, evaluation, and feedback can help identify and address biases in real-time, thereby enabling continuous improvement and refinement of AI systems over time.

Conclusion

Mitigating bias in AI systems is paramount for ensuring the security, fairness, and effectiveness of web3 companies' operations. By addressing native biases, conditional biases, and bias factors arising from biases against bias, organizations can enhance the reliability and trustworthiness of their AI-driven solutions. It is essential to recognize that while bias mitigation is crucial, it must be approached with nuance and context sensitivity to avoid inadvertently suppressing valid factors and compromising the utility of AI algorithms.

Furthermore, achieving effective bias mitigation requires a multidisciplinary approach that integrates technical

[571] *Id.*

expertise with ethical considerations, legal frameworks, and stakeholder engagement. Collaboration among data scientists, ethicists, policymakers, and domain experts is essential for developing comprehensive strategies that balance the imperatives of fairness, accuracy, and utility in AI systems. Transparency, accountability, and continuous monitoring are key principles that should guide bias mitigation efforts, enabling organizations to identify and address biases proactively while fostering trust and transparency with stakeholders.

As AI technologies continue to evolve and proliferate, ongoing research, innovation, and collaboration will be essential for advancing the state of the art in bias mitigation and promoting responsible AI development practices. By embracing diversity, equity, and inclusion principles throughout the AI lifecycle, companies can harness the full potential of AI to drive innovation, enhance decision-making, and deliver value to customers while mitigating the risks of unintended consequences and discriminatory outcomes.

By prioritizing bias mitigation in AI development and deployment, web3 companies can uphold ethical standards, comply with regulatory requirements, and build trust with users and communities. Embracing a proactive and holistic approach to bias mitigation will not only safeguard against security vulnerabilities but also contribute to the creation of a more equitable and inclusive digital ecosystem in which AI serves as a force for positive societal impact.

Chapter 32: Community Engagement

"Automation is no longer just a problem for those working in manufacturing. Physical labor was replaced by robots; mental labor is going to be replaced by AI and software."

Andrew Yang[572]

This chpater explores the crucial role of community engagement in ensuring the security of web3 companies. Effective community engagement serves as a cornerstone for building trust, fostering collaboration, and enhancing cybersecurity resilience. In today's interconnected digital landscape, where data breaches and cyber threats are ever-present risks, establishing meaningful relationships with stakeholders is paramount for safeguarding sensitive information and mitigating potential security vulnerabilities. This chapter delves into the importance of community identification, understanding the information needs of different stakeholders, and implementing strategies to maintain an appropriate balance between transparency and confidentiality.

Community identification is the initial step in effective engagement efforts, requiring companies to identify and understand the diverse groups that comprise their community. This includes not only customers and clients but also employees, partners, regulators, and other relevant stakeholders. By delineating the various segments of their community, companies can tailor their engagement strategies to address the specific needs, concerns, and expectations of each group. Moreover, community identification enables organizations to proactively identify

[572] (2024). "AI Quotes, Andrew Yang." Brainy Quote at https://www.brainyquote.com/topics/ai-quotes.

potential security risks and vulnerabilities, allowing for targeted interventions and risk mitigation measures.

Understanding the information needs of the community is essential for providing relevant, timely, and accurate communication that fosters trust and transparency. Companies must discern what information is essential for stakeholders to make informed decisions and what information may pose security risks if disclosed. Striking the right balance between transparency and confidentiality requires careful consideration of factors such as data sensitivity, regulatory requirements, and stakeholder preferences. By aligning their communication strategies with the information needs of the community, companies can enhance their credibility and reputation while minimizing the likelihood of security incidents stemming from inadequate or excessive disclosure.

Conversely, failing to provide essential information to the community can undermine trust and lead to speculation, rumors, and misinformation, creating fertile ground for security threats and social engineering attacks. Therefore, companies must adopt proactive communication practices that keep stakeholders informed about relevant security policies, procedures, and incidents. Transparency builds confidence and empowers the community to actively participate in security efforts, such as reporting suspicious activities or vulnerabilities, thereby strengthening the overall cybersecurity posture of the organization.

At the same time, it is crucial to ensure that AI systems deployed for community engagement purposes do not inadvertently divulge sensitive information or compromise privacy. AI algorithms must be carefully designed and trained to balance the need for personalized communication with the imperative of data protection. Robust data anonymization techniques, encryption protocols, and access controls can help mitigate the risks associated with AI-driven communication while preserving the confidentiality and integrity of sensitive information. By integrating security-by-design principles into AI development processes, companies can leverage technology to engage with their community effectively while upholding privacy and security standards.

Who is your community?

Identifying the community is a foundational step in enhancing digital security for web3 companies. By clearly defining the various stakeholders that constitute their community, organizations can develop targeted security strategies tailored to the specific needs, concerns, and expectations of each group.[573] This process involves identifying not only external stakeholders such as customers, clients, and partners but also internal stakeholders such as employees, executives, and shareholders. Understanding the composition of the community allows companies to prioritize their security efforts and allocate resources effectively to address the most critical areas of concern.

Furthermore, community identification enables companies to assess the potential security risks associated with each stakeholder group and develop mitigation strategies accordingly. Different segments of the community may have varying levels of access to sensitive information, varying degrees of technological proficiency, and different threat profiles.[574] By categorizing stakeholders based on factors such as role, access privileges, and interaction

---

[573] Smithmyer et al, *supra*.
[574] *Id.*

patterns, organizations can implement targeted security measures to protect against insider threats, social engineering attacks, and other cyber risks specific to each group.

In addition to mitigating security risks, identifying the community facilitates effective communication and collaboration on security-related matters. By understanding the unique needs and preferences of stakeholders, companies can tailor their communication strategies to deliver relevant security information, raise awareness about potential threats, and promote best practices for cybersecurity.[575] Engaging with the community in a transparent and proactive manner fosters trust and confidence, encouraging stakeholders to actively participate in security initiatives, report incidents, and adhere to security policies and procedures.

Community identification plays a crucial role in crisis management and incident response. In the event of a security breach or cyber attack, having a clear understanding of the affected stakeholders enables companies to prioritize communication efforts, provide timely updates, and offer support and assistance where needed. By maintaining open channels of communication with the community, organizations can mitigate the reputational damage and financial losses associated with security incidents, demonstrating their commitment to transparency, accountability, and customer-centric security practices.

---

[575] *Id.*

Information your community needs

Identifying the information needs of the community is essential for ensuring effective security measures within a web3 company. By understanding the specific data requirements of different stakeholders, organizations can tailor their security strategies to prioritize the protection of sensitive information while also facilitating the legitimate access and sharing of relevant data. This process involves conducting thorough assessments to determine the types of information that are critical to each segment of the community and the purposes for which it is needed.

Identifying the information needs of the community allows companies to establish clear policies and procedures governing the collection, storage, transmission, and disposal of data. By defining what constitutes sensitive information and establishing guidelines for its handling, organizations can ensure compliance with relevant privacy regulations and industry standards while also safeguarding against unauthorized access, disclosure, or misuse.[576] Effective data governance practices help mitigate the risk of data breaches, identity theft, and other security incidents that could compromise the trust and confidence of stakeholders.

Furthermore, understanding the information needs of the community enables companies to implement appropriate access controls and authentication mechanisms to restrict access to sensitive data only to authorized individuals or entities.[577] By adopting a risk-based approach to access management, organizations can balance the need for data security with the requirements for operational efficiency

[576] Smithmyer, *supra.*
[577] *Id.*

and user convenience. This involves implementing granular access controls, role-based permissions, and multi-factor authentication mechanisms to ensure that sensitive information is protected from unauthorized access or misuse while still enabling legitimate users to perform their roles effectively.

Additionally, identifying the information needs of the community facilitates effective communication and engagement on security-related issues. By providing stakeholders with access to relevant information, resources, and support channels, organizations can empower them to take proactive steps to protect themselves against cyber threats, such as phishing attacks, malware infections, and social engineering scams.[578] Transparent communication builds trust and confidence among stakeholders, fostering a culture of security awareness and collaboration that enhances the overall resilience of the organization against evolving cyber risks.

Information your community does not need.

Identifying the information that the community does not need and should not have access to is critical for maintaining robust security measures within a web3 company.[579] While it is important to provide stakeholders with relevant data to support their roles and responsibilities, allowing access to unnecessary or sensitive information can increase the risk of data breaches, privacy violations, and other security incidents. Therefore, organizations must carefully assess and identify the types of information that should be restricted or withheld from certain individuals or groups within the community.

[578] *Id.*
[579] Smithmyer et al, *supra*.

First, limiting access to unnecessary information helps minimize the attack surface and reduce the likelihood of unauthorized access or exploitation by malicious actors. By implementing the principle of least privilege, organizations can restrict access to sensitive data only to individuals who require it to perform their job functions effectively.[580] This approach helps mitigate the risk of insider threats, accidental data leaks, and other security breaches that could compromise the confidentiality, integrity, or availability of critical information.

Second, identifying the information that the community does not need helps streamline data management processes and improve overall data hygiene. By removing unnecessary data from systems and repositories, organizations can reduce the complexity and overhead associated with data storage, processing, and maintenance. This simplification not only enhances operational efficiency but also minimizes the risk of data sprawl, data redundancy, and data retention compliance issues that could pose security and compliance challenges for the organization.

Thirdly, restricting access to irrelevant or sensitive information helps protect the privacy and confidentiality of individuals' personal data.[581] In compliance with data protection regulations such as the General Data Protection Regulation (GDPR) and the California Consumer Privacy Act (CCPA), organizations have a legal obligation to safeguard sensitive personal information from unauthorized access, disclosure, or misuse. By identifying and categorizing sensitive data, companies can implement appropriate access controls, encryption measures, and data

---

[580] *Id.*
[581] Smithmyer, *supra.*

anonymization techniques to protect individuals' privacy rights and prevent unauthorized data processing activities.

Limiting access to unnecessary information reduces the risk of information overload and cognitive overload among stakeholders, which can hinder productivity and decision-making processes. By focusing on providing relevant, actionable information to users, organizations can enhance user experience and satisfaction while minimizing the likelihood of errors, misunderstandings, or misinterpretations that could arise from information overload.[582] This targeted approach to information delivery ensures that users can quickly access the data they need to perform their tasks efficiently without being overwhelmed by irrelevant or extraneous information.

Finally, identifying the information that the community does not need fosters a culture of data stewardship and responsible information management within the organization. By promoting awareness of data privacy and security best practices, organizations can empower employees to make informed decisions about data handling and usage, reducing the risk of data breaches or compliance violations. This proactive approach to data governance strengthens the organization's overall security posture and enhances its reputation as a trusted custodian of sensitive information.

Ensuring your AI does not give out too much information

Restricting AI to ensure that it provides the community with an appropriate amount of information involves a delicate balance between transparency, privacy, and security considerations. While there are benefits to both

---

[582] *Id.*

providing too much and too little information, there are also potential drawbacks and challenges associated with each approach.

On one hand, providing too much information can lead to information overload, confusion, and cognitive overload among community members. When AI systems inundate users with excessive data, it can be challenging for individuals to identify relevant information and make informed decisions.[583] Moreover, the indiscriminate sharing of vast amounts of data may inadvertently expose sensitive information or proprietary knowledge, increasing the risk of data breaches, privacy violations, and intellectual property theft. Therefore, restricting AI to avoid overwhelming users with unnecessary or irrelevant information can help streamline communication, improve user experience, and protect sensitive data.

However, overly restrictive AI systems may also limit access to valuable insights, hinder collaboration, and stifle innovation within the community. When AI algorithms are overly cautious in withholding information, they may inadvertently withhold critical data or insights that could drive meaningful discussions, inform decision-making processes, or spark creative solutions.[584] Moreover, excessive restrictions on information sharing can erode trust and transparency within the community, leading to skepticism, frustration, and disengagement among users. Therefore, it is essential to strike a balance between providing users with access to relevant information and

---

[583] Hamet, P. and Tremblay, J., 2017. Artificial intelligence in medicine. *Metabolism*, *69*, pp.S36-S40.
[584] Janssen, M., Brous, P., Estevez, E., Barbosa, L.S. and Janowski, T., 2020. Data governance: Organizing data for trustworthy Artificial Intelligence. *Government information quarterly*, *37*(3), p.101493.

safeguarding sensitive or proprietary data to promote openness, collaboration, and trust within the community.

Furthermore, the level of information provided by AI systems must be tailored to meet the diverse needs, preferences, and expertise levels of different users within the community. While some users may require detailed, granular information to perform their roles effectively, others may prefer concise summaries or high-level overviews to facilitate understanding and decision-making. Therefore, AI systems should incorporate user-centric design principles and customizable features to enable users to adjust the level of information granularity according to their preferences and requirements. By empowering users to personalize their information consumption experience, organizations can enhance user satisfaction, productivity, and engagement while maintaining data privacy and security.

AI systems must adhere to ethical and regulatory guidelines governing data privacy, confidentiality, and transparency to ensure responsible information disclosure and usage. When determining the appropriate level of information to provide, AI algorithms should prioritize user privacy, consent, and data protection principles to prevent unauthorized access, disclosure, or misuse of sensitive information.[585] Additionally, organizations must implement robust data governance frameworks, security controls, and audit mechanisms to monitor and mitigate the risks associated with information sharing and access.

[585] Janssen, M., Brous, P., Estevez, E., Barbosa, L.S. and Janowski, T., 2020. Data governance: Organizing data for trustworthy Artificial Intelligence. *Government information quarterly*, *37*(3), p.101493.

However, overly restrictive AI systems may also hinder knowledge sharing, collaboration, and community engagement by limiting access to valuable insights, expertise, and resources. When AI algorithms err on the side of caution by withholding information, they may inadvertently impede the flow of information, innovation, and collaboration within the community. Moreover, excessive restrictions on information sharing can breed mistrust, skepticism, and frustration among users, undermining the credibility and effectiveness of AI-powered systems. Therefore, organizations must carefully balance the need for transparency and accountability with the imperative to protect sensitive data and intellectual property rights to foster a culture of trust, collaboration, and innovation within the community.

In the end, achieving the right balance between providing too much and too little information requires organizations to adopt a proactive, user-centric approach to AI design, implementation, and governance. By leveraging advanced AI technologies, user feedback, and iterative design methodologies, organizations can develop AI systems that deliver timely, relevant information to users while safeguarding privacy, security, and ethical considerations. Organizations must cultivate a culture of transparency, openness, and accountability to ensure that AI systems serve the best interests of the community while upholding the highest standards of integrity and ethical conduct. Through continuous monitoring, evaluation, and refinement, organizations can optimize the performance, usability, and trustworthiness of AI-powered systems to empower users, foster collaboration, and drive positive outcomes within the community.

Conclusion

Part VII delves into the complex interplay between AI, machine learning, and security, highlighting the critical importance of transparency, accountability, and responsible innovation in mitigating risks and safeguarding digital ecosystems. From addressing algorithmic bias to ensuring community engagement and data privacy, the chapters underscore the multifaceted challenges and opportunities inherent in leveraging AI technologies for security purposes. By examining the nuances of algorithmic decision-making, bias mitigation, and community interaction, organizations can better understand the implications of AI for security and devise effective strategies to navigate the evolving landscape of digital risk.

Throughout Part VII, the discussions emphasize the need for proactive measures to address algorithmic bias, discrimination, and privacy concerns in AI systems. By implementing robust transparency mechanisms, accountability frameworks, and bias mitigation strategies, organizations can mitigate the risks of unintended consequences and uphold ethical standards in AI development and deployment. Moreover, fostering collaboration, inclusivity, and community engagement is paramount to building trust, credibility, and resilience in AI-powered security solutions. By involving stakeholders in the design, implementation, and evaluation of AI systems, organizations can enhance transparency, accountability, and user satisfaction while minimizing the potential for bias, discrimination, and privacy violations.

Furthermore, Part VII underscores the importance of aligning AI practices with regulatory requirements, ethical principles, and industry standards to promote responsible innovation and ensure compliance with legal and ethical obligations. By adhering to best practices in AI governance, data protection, and risk management, organizations can mitigate the risks of AI-related security breaches, data leaks, and reputational damage. Moreover, investing in ongoing training, education, and awareness initiatives is essential to promoting a culture of cybersecurity, privacy, and ethical AI use across the organization.

In summation, Part VII provides valuable insights into the complex challenges and opportunities associated with AI and machine learning in the context of security. By embracing transparency, accountability, and responsible innovation, organizations can harness the transformative potential of AI while mitigating risks, enhancing trust, and safeguarding digital ecosystems. As AI continues to reshape the landscape of security and risk management, organizations must remain vigilant, adaptable, and ethically conscious to navigate the evolving complexities of the digital age effectively.

Chapter 33: Investment and Development Monitoring

"Humans have 3 percent human error, and a lot of companies can't afford to be wrong 3 percent of the time anymore, so we close that 3 percent gap with some of the technologies. The AI we've developed doesn't make mistakes."

Walter O'Brian[586]

Chapter 33 jumps into the crucial aspect of investment and development monitoring within the realm of AI and machine learning security. In today's digital world, organizations are increasingly relying on investment bots and sophisticated algorithms to manage their investment portfolios, drive development initiatives, and optimize supply chain management. However, with the growing complexity of AI-driven investment strategies and development projects, it becomes imperative for companies to implement robust monitoring mechanisms to track investment performance, assess developmental progress, and ensure supply chain integrity.

One of the key areas explored in this chapter is the use of investment bots, automated trading algorithms, and AI-driven investment strategies in managing financial portfolios. Investment bots leverage machine learning algorithms to analyze market trends, identify lucrative investment opportunities, and execute trades in real-time. However, the effectiveness and reliability of these bots hinge on their ability to generate favorable investment

[586] (2024). "AI Quotes, Walter O'Brian." Brainy Quote at https://www.brainyquote.com/topics/ai-quotes.

results while minimizing risks and adhering to predefined investment objectives.

The chapter delves into the importance of investment results monitoring, which involves tracking and evaluating the performance of investment portfolios over time. By leveraging AI-powered analytics and data visualization tools, organizations can gain valuable insights into investment trends, portfolio volatility, and risk exposure. Additionally, continuous monitoring enables stakeholders to make informed decisions, adjust investment strategies, and optimize portfolio allocation to maximize returns and mitigate financial risks.

In addition to investment monitoring, the chapter examines the role of portfolio monitoring in ensuring the overall health and performance of investment portfolios. Portfolio monitoring encompasses assessing the composition, diversification, and risk profile of investment portfolios to align them with strategic objectives and risk tolerance levels. Through proactive monitoring and analysis, organizations can identify portfolio imbalances, asset allocation discrepancies, and potential areas of vulnerability, allowing them to take timely corrective actions and optimize portfolio performance.

Furthermore, the chapter explores the importance of developmental planning and results monitoring in driving innovation and managing development projects effectively. Developmental planning involves defining project goals, milestones, and resource allocation strategies to guide the execution of development initiatives. Meanwhile, developmental results monitoring entails tracking project progress, evaluating performance metrics, and identifying areas for improvement to ensure that development projects stay on track and deliver desired outcomes. By leveraging

AI-driven analytics and project management tools, organizations can enhance visibility, transparency, and accountability in the development process, thereby minimizing risks and maximizing project success.

Investment Bots

Investment bots play a crucial role in the financial management of web3 companies, offering several advantages in terms of efficiency, accuracy, and automation. These sophisticated algorithms leverage machine learning and artificial intelligence to analyze vast amounts of market data, identify investment opportunities, and execute trades with precision and speed.[587] For web3 companies operating in dynamic and volatile markets, investment bots provide a valuable tool for optimizing investment decisions and maximizing returns on investment.

One significant advantage of investment bots is their ability to process large volumes of data and execute trades with minimal human intervention. By leveraging advanced analytics and predictive modeling techniques, these bots can quickly identify market trends, patterns, and anomalies, allowing web3 companies to capitalize on lucrative investment opportunities in real-time.[588] Moreover, investment bots can operate 24/7, continuously monitoring market conditions and executing trades according to predefined parameters, thereby eliminating human error and ensuring consistent performance.

---

[587] Khurana, S.S., Singh, P. and Garg, N.K., 2023. Revolutionize AI Trading Bots with AutoML-Based Multi-timeframe Bitcoin Price Prediction. *SN Computer Science*, *4*(5), p.492.
[588] *Id.*

Furthermore, investment bots offer web3 companies the advantage of speed and agility in executing trades, enabling them to capitalize on fleeting market opportunities and respond swiftly to changing market conditions. With the ability to execute trades within milliseconds, these bots can exploit price differentials and arbitrage opportunities across multiple markets, generating profits for the company in a highly competitive landscape. Additionally, investment bots can automate the execution of complex trading strategies, such as algorithmic trading, quantitative analysis, and high-frequency trading, allowing web3 companies to implement sophisticated investment strategies with precision and efficiency.

Despite their benefits, investment bots also pose certain drawbacks and risks for web3 companies, which must be carefully considered and managed. One major concern is the potential for algorithmic errors or glitches, which can lead to unexpected losses or disruptions in trading operations. Given the complexity of AI-driven algorithms, even minor errors or bugs can have significant consequences, resulting in financial losses or reputational damage for the company.[589] Moreover, investment bots are susceptible to market volatility and unpredictable events, which can exacerbate risks and lead to unexpected outcomes.

Another drawback of investment bots is the risk of overreliance on automation and the erosion of human oversight and judgment in investment decision-making. While these bots excel at processing data and executing trades with speed and efficiency, they lack the intuition,

---

[589] Jazayeriy, H. and Daryani, M., 2021, December. Spa bot: Smart price-action trading bot for cryptocurrency market. In *2021 12th International Conference on Information and Knowledge Technology (IKT)* (pp. 178-182). IEEE.

experience, and contextual understanding that human traders bring to the table. As a result, web3 companies may become overly dependent on algorithmic trading strategies, neglecting the importance of human oversight and risk management in investment operations.

Additionally, investment bots can contribute to market inefficiencies and create liquidity issues, particularly in highly automated and algorithm-driven markets. The rapid proliferation of investment bots and algorithmic trading strategies can amplify market volatility, exacerbate price fluctuations, and distort market dynamics, leading to adverse consequences for market participants and investors. Moreover, the use of investment bots by large institutional players and hedge funds can exacerbate inequalities and exacerbate market manipulation, posing systemic risks to the financial system.

While investment bots offer web3 companies significant benefits in terms of efficiency, speed, and automation, they also present inherent risks and challenges that must be carefully managed. By understanding the limitations and drawbacks of investment bots and implementing robust risk management practices, web3 companies can leverage these tools effectively to optimize their investment strategies and achieve their financial objectives while mitigating potential risks and vulnerabilities.

Investment Results Monitoring

AI plays a pivotal role in investment results monitoring for Web3 companies, offering a range of benefits that enhance decision-making processes and optimize portfolio performance. One key advantage of AI in investment results monitoring is its ability to analyze vast amounts of financial data with speed and accuracy, providing real-time insights into market trends, portfolio performance, and investment opportunities. By leveraging advanced machine learning algorithms, AI systems can identify patterns, anomalies, and correlations in financial data, enabling web3 companies to make informed investment decisions and adjust their portfolios dynamically in response to changing market conditions.

Furthermore, AI-driven investment results monitoring enables web3 companies to gain deeper insights into portfolio performance metrics, risk exposures, and asset allocation strategies. Through sophisticated data analytics and predictive modeling techniques, AI systems can assess the performance of individual investments, evaluate portfolio diversification strategies, and optimize risk-adjusted returns.[590] By analyzing historical data and forecasting future market trends, AI algorithms empower web3 companies to identify potential risks and opportunities proactively, thereby enhancing the overall effectiveness of their investment strategies.

AI enhances investment results monitoring by providing customizable dashboards, interactive visualizations, and personalized insights that enable web3 companies to monitor and evaluate their portfolios in real-time. By aggregating and synthesizing data from multiple sources, AI-powered dashboards offer a comprehensive view of portfolio performance metrics, investment returns, and key performance indicators, allowing stakeholders to track progress, identify areas for improvement, and make data-driven decisions with confidence.

---

[590] Smith, K.J., 1993. Investment Monitoring Systems, Abandonment of Capital Assets, and Firm Performance. *Journal of Management Accounting Research, 5.*

Despite its numerous benefits, AI-driven investment results monitoring also presents certain drawbacks and challenges that must be addressed to ensure its effectiveness and reliability. One major concern is the potential for algorithmic bias and model drift, which can lead to inaccurate predictions, flawed investment decisions, and unexpected losses. To mitigate these risks, web3 companies must implement robust validation processes, model governance frameworks, and algorithmic transparency measures to ensure the integrity and fairness of AI-driven investment models.

Additionally, AI-powered investment results monitoring may face challenges related to data quality, data privacy, and data security, particularly when dealing with sensitive financial information and proprietary data sets. Web3 companies must adopt stringent data governance policies, data encryption techniques, and cybersecurity protocols to safeguard against data breaches, unauthorized access, and malicious attacks that could compromise the confidentiality, integrity, and availability of their investment data.

AI-driven investment results monitoring may raise ethical and regulatory concerns related to algorithmic accountability, transparency, and regulatory compliance. As AI systems become increasingly sophisticated and autonomous, questions arise about the ethical implications of algorithmic decision-making, the potential for unintended consequences, and the need for regulatory oversight to ensure fairness, transparency, and accountability in financial markets.

AI offers significant advantages in investment results monitoring for web3 companies, enabling them to analyze data more effectively, make better-informed investment decisions, and optimize portfolio performance. However, web3 companies must also address the challenges and risks associated with AI-driven investment monitoring, including algorithmic bias, data privacy concerns, and regulatory compliance issues, to maximize the benefits of AI while mitigating potential drawbacks and vulnerabilities.

Portfolio Monitoring

Beyond monitoring investment results, AI can play a crucial role in analyzing various other analytic variables within an investment portfolio, thereby enhancing investment security and performance. One key area where AI excels is in evaluating risk factors associated with individual investments, asset classes, and market segments. By leveraging sophisticated machine learning algorithms, AI systems can assess historical performance data, market volatility, macroeconomic indicators, and other relevant factors to identify potential risks and opportunities, enabling investors to make more informed decisions and mitigate downside risks effectively.

Furthermore, AI-driven portfolio analytics can provide valuable insights into portfolio diversification strategies, asset allocation optimization, and risk management techniques. By analyzing correlations, covariance matrices, and other statistical measures, AI algorithms can help investors build well-balanced portfolios that are resilient to market fluctuations and economic downturns.[591] Additionally, AI-powered analytics can identify emerging trends, sector rotations, and investment themes, enabling investors to capitalize on opportunities and adjust their

[591] Marinho, P., Fernandes, P. and Pimentel, F., 2021. The digital portfolio as an assessment strategy for learning in higher education. *Distance Education*, *42*(2), pp.253-267.

portfolios dynamically in response to changing market conditions, thereby enhancing investment security and performance over time.

AI can assist investors in optimizing their investment strategies by providing personalized recommendations, predictive insights, and scenario analysis tools. By leveraging advanced data analytics and predictive modeling techniques, AI systems can forecast future market trends, simulate hypothetical scenarios, and recommend optimal investment strategies tailored to each investor's risk tolerance, financial goals, and investment preferences. This personalized approach to investment management not only enhances investment security by aligning investment decisions with investors' objectives but also improves overall portfolio performance by maximizing returns and minimizing risks.

Developmental Planning

AI can significantly enhance security in establishing protocols for secure product development procedures, particularly in preventing leaks of new products and intellectual property. One of the primary ways AI contributes to this aspect of security is through its ability to monitor and analyze vast amounts of data generated during the product development lifecycle. By deploying AI-powered algorithms to scrutinize internal communications, project management systems, and collaboration platforms, companies can detect any unauthorized access or suspicious activities that may indicate potential leaks or breaches of confidentiality.

Furthermore, AI-driven anomaly detection systems can identify deviations from normal behavior patterns among employees, contractors, or third-party vendors involved in the product development process. These systems can flag unusual access patterns, unauthorized file transfers, or attempts to access sensitive information outside of designated workflows, enabling security teams to investigate and mitigate potential threats proactively. Additionally, AI algorithms can analyze metadata associated with documents, code repositories, and project files to identify any discrepancies or inconsistencies that may signal unauthorized modifications or tampering attempts.

Another way AI enhances security in product development procedures is by automating compliance checks and adherence to security protocols throughout the development lifecycle. By integrating AI-powered compliance monitoring tools into development environments, companies can ensure that all stakeholders adhere to established security policies, regulatory requirements, and best practices. These tools can automatically scan code repositories, development environments, and deployment pipelines for vulnerabilities, misconfigurations, or non-compliant activities, thereby reducing the risk of security incidents and ensuring the integrity of the development process.

AI-driven threat intelligence platforms can provide real-time insights into emerging cyber threats, vulnerabilities, and attack vectors that may pose risks to product development initiatives. By aggregating and analyzing data from various sources, including security feeds, dark web forums, and open-source intelligence, these platforms can identify potential threats targeting specific industries, technologies, or geographic regions. Armed with this intelligence, companies can proactively adjust their security posture, update their defenses, and implement additional safeguards to protect their intellectual property and sensitive information from cyber adversaries.

Additionally, AI can play a crucial role in ensuring the security of collaboration and communication channels used during product development, such as email, messaging platforms, and video conferencing tools. AI-powered email security solutions can analyze incoming and outgoing messages for signs of phishing attempts, malware attachments, or social engineering attacks, thereby reducing the risk of unauthorized access or data breaches. Similarly, AI-driven data loss prevention (DLP) systems can monitor file sharing activities, enforce encryption policies, and prevent the unauthorized exfiltration of sensitive information, safeguarding valuable intellectual property throughout the development process.

Developmental Results Monitoring

AI can significantly bolster security in product development procedures by establishing robust protocols that prevent leaks of sensitive information regarding new products. One of the primary ways AI achieves this is through the implementation of advanced data encryption and access control mechanisms, which safeguard proprietary data from unauthorized access or disclosure. By employing encryption algorithms and multi-factor authentication protocols, AI-powered systems can ensure that only authorized personnel with the requisite permissions can access confidential project documents, designs, and specifications, thereby minimizing the risk of data breaches or intellectual property theft.[592]

[592] *Id.*

Furthermore, AI-driven monitoring and surveillance tools can continuously monitor internal networks, communication channels, and file repositories for any suspicious or unauthorized activities. By leveraging machine learning algorithms to analyze patterns, anomalies, and user behaviors, AI systems can flag potential security breaches in real-time, allowing security teams to respond promptly and mitigate any threats before they escalate. Moreover, AI-powered digital rights management (DRM) solutions can track the usage and distribution of sensitive documents, detect unauthorized sharing or duplication attempts, and enforce strict access controls to prevent data leaks or unauthorized disclosures.

Additionally, AI can enhance security by automating the enforcement of compliance policies, regulatory standards, and industry best practices throughout the product development lifecycle. By embedding compliance checks, audit trails, and regulatory frameworks into AI-driven workflow automation systems, companies can ensure that all development activities adhere to legal and regulatory requirements, thereby reducing the risk of non-compliance penalties, legal disputes, or reputational damage.[593] AI-powered risk assessment tools can proactively identify potential security vulnerabilities, design flaws, or compliance gaps early in the development process, allowing companies to address them promptly and prevent costly security incidents or product recalls down the line.

Supply Chain Management

AI can play a crucial role in enhancing security in supply chain management by offering advanced analytics and predictive capabilities. Firstly, AI can be used to analyze

---

[593] *Id.*

historical data and detect patterns of fraudulent activities or potential risks within the supply chain. By identifying anomalies or irregularities in transactions, shipments, or supplier behavior, AI algorithms can alert companies to possible security threats, such as counterfeit products, unauthorized access, or supplier fraud.[594]

AI-driven predictive analytics can help forecast demand, optimize inventory levels, and improve overall supply chain efficiency. By accurately predicting demand fluctuations and supply chain disruptions, companies can proactively address potential security vulnerabilities and minimize the risk of stockouts, delays, or overstock situations, which could lead to financial losses or reputational damage.

Furthermore, AI can facilitate real-time monitoring and tracking of shipments and goods throughout the supply chain, enabling companies to ensure the integrity and security of their products. Through the use of IoT sensors, RFID tags, and blockchain technology, AI-powered systems can provide end-to-end visibility and transparency, allowing companies to trace the movement of goods, verify authenticity, and detect any unauthorized tampering or theft.

Additionally, AI-driven risk assessment tools can evaluate supplier performance, assess compliance with regulatory standards, and identify high-risk suppliers or regions prone to security breaches. By conducting automated risk assessments and due diligence checks, companies can make informed decisions when selecting suppliers and mitigate

---

[594] Helo, P. and Hao, Y., 2022. Artificial intelligence in operations management and supply chain management: An exploratory case study. *Production Planning & Control*, *33*(16), pp.1573-1590..

potential security risks associated with unreliable or non-compliant partners.

AI can enable proactive risk mitigation strategies by continuously monitoring external factors, such as geopolitical events, market trends, or environmental disruptions, that may impact the supply chain.[595] By analyzing vast amounts of data from various sources in real-time, AI systems can provide early warnings and recommend preventive measures to safeguard the supply chain against potential security threats or disruptions.

Conclusion

AI's utility extends beyond merely monitoring investment results; it can delve into various analytic variables within an investment portfolio, bolstering investment security and performance. AI's prowess shines in evaluating risk factors associated with investments, asset classes, and market segments. Leveraging sophisticated algorithms, it sifts through historical data, market volatility, and macroeconomic indicators to pinpoint risks and opportunities, empowering investors to make informed decisions and effectively mitigate downside risks.

AI-driven portfolio analytics offer insights into diversification strategies, asset allocation optimization, and risk management techniques. By scrutinizing correlations and statistical measures, AI aids in constructing well-balanced portfolios resilient to market fluctuations. It identifies emerging trends, enabling investors to capitalize on opportunities and dynamically adjust their portfolios. This dynamic approach enhances investment security and

[595] *Id.*

performance over time, ensuring portfolios remain agile and responsive to evolving market conditions.

Additionally, AI assists in optimizing investment strategies through personalized recommendations, predictive insights, and scenario analysis. By harnessing data analytics and predictive modeling, AI forecasts market trends, simulates scenarios, and tailors investment strategies to individual risk tolerance and financial goals. This personalized approach aligns investment decisions with investors' objectives, enhancing investment security, and maximizing returns. Thus, AI's multifaceted analytics not only fortify investment security but also elevate portfolio performance, ushering in a new era of data-driven investment management.

Conclusion Part VII

Part VII of our exploration into security in the digital environment has been a voyage through the intricate realms of AI and machine learning, focusing on their application in investment and development monitoring within web3 companies. From the foundational understanding of cybersecurity in Chapter 30 to the nuanced examination of investment strategies and portfolio management in Chapter 33, each chapter has contributed vital insights into the evolving landscape of digital security.

Chapter 30 laid the groundwork by elucidating the fundamentals of cybersecurity, emphasizing the importance of robust defense mechanisms and proactive risk management strategies in safeguarding digital assets. It provided a comprehensive overview of cyber threats, vulnerabilities, and best practices, serving as a primer for the subsequent chapters' discussions.

Building upon this foundation, Chapter 31 delved into the intricacies of data privacy and compliance in the digital age, shedding light on the regulatory landscape and ethical considerations surrounding data protection. It underscored the imperative for web3 companies to uphold stringent data privacy standards and navigate regulatory frameworks to foster trust and mitigate legal risks.

Chapter 32 marked a pivotal juncture in our exploration, exploring the convergence of AI and cybersecurity and its implications for threat detection, incident response, and vulnerability management. It showcased the transformative potential of AI-driven technologies in augmenting cybersecurity defenses, enhancing threat intelligence capabilities, and enabling proactive defense strategies.

Finally, Chapter 33 brought our journey to a crescendo by delving into the critical domain of investment and development monitoring within the realm of AI and machine learning security. It underscored the indispensable role of investment bots, automated trading algorithms, and AI-driven analytics in optimizing financial management practices and driving innovation within web3 companies.

Collectively, these chapters have illuminated the multifaceted nature of security in the digital environment, emphasizing the symbiotic relationship between technological advancements and cybersecurity strategies. They have underscored the need for web3 companies to embrace AI and machine learning technologies responsibly, leveraging their transformative potential while mitigating risks and ensuring ethical considerations.

Part VIII: Security on the Global Level

Introduction Part VIII

Part VIII embarks on a profound exploration into the dynamic landscape of emerging technologies, ethical considerations, and societal impacts in the ever-evolving realm of web3. As the digital sphere continues to evolve at a rapid pace, Part VIII delves into the forefront of innovation and disruption, presenting a compelling narrative that transcends conventional boundaries and redefines the future of technology and human interaction.

In this introductory chapter, we set the stage for an enlightening journey through the transformative potential and ethical implications of emerging technologies in the web3 era. From artificial intelligence and blockchain to augmented reality and quantum computing, the chapters ahead will explore the myriad ways in which these technologies are reshaping industries, revolutionizing business models, and redefining the human experience.

At the heart of Part VIII lies a deep-seated commitment to exploring the ethical dimensions of technological advancement and the societal impact of innovation. As we navigate the complex terrain of ethical dilemmas, privacy concerns, and social responsibility, we strive to foster critical dialogue and thoughtful reflection on the implications of our digital future.

Through a multidisciplinary lens, Part VIII endeavors to unravel the intricate interplay between technology, ethics, and society, offering insights from experts across diverse fields and perspectives. From scholars and policymakers to industry leaders and innovators, our contributors bring a wealth of knowledge and expertise to bear on the pressing issues and emerging trends shaping the digital landscape.

Part VIII seeks to inspire meaningful action and responsible stewardship in the face of technological disruption and societal change. By illuminating the ethical considerations and societal impacts of emerging technologies, we aim to empower individuals and organizations to navigate the complexities of the digital age with integrity, empathy, and foresight.

In the chapters ahead, readers will encounter thought-provoking discussions on topics ranging from algorithmic bias and data privacy to digital rights and corporate responsibility. Through engaging narratives, case studies, and expert analysis, Part VIII invites readers to explore the transformative potential of emerging technologies while grappling with the ethical and moral challenges they pose.

As we embark on this intellectual journey, let us approach the exploration of emerging technologies with curiosity, humility, and a commitment to fostering a more inclusive, equitable, and ethical digital future. Together, let us navigate the complexities of the web3 era with wisdom, compassion, and a steadfast dedication to the greater good.

Chapter 34: Regulatory Landscape

"If you have 10,000 regulations you destroy all respect for the Law"

Winston Churchill[596]

Chapter 34 shines a spotlight on the regulatory landscape's formidable challenges, particularly in the domains of privacy regulations and data security. With the proliferation of data breaches and privacy concerns, governments worldwide have enacted stringent regulations to safeguard individuals' privacy rights and protect sensitive data from unauthorized access and misuse. From the European Union's General Data Protection Regulation (GDPR) to the California Consumer Privacy Act (CCPA), web3 companies grapple with compliance requirements, data protection obligations, and the need for robust cybersecurity measures to mitigate risks and maintain regulatory compliance.

Furthermore, Chapter 34 delves into the intricacies of regulations pertaining to turning over information related to warrants. In an age where law enforcement agencies seek access to digital evidence for investigative purposes, web3 companies must navigate a complex legal landscape governing information disclosure and user privacy rights. Balancing the imperatives of law enforcement with the protection of user privacy poses significant challenges for web3 companies, requiring careful consideration of legal obligations, transparency requirements, and the ethical implications of information disclosure.

[596] (2024). "Regulation Quotes, Winston Churchill," Brainy Quote at https://www.brainyquote.com/topics/regulations-quotes.

The chapter explores regulations governing publicly disclosed information and their implications for web3 companies operating in transparent and accountable digital ecosystems. As governments and regulatory bodies mandate greater transparency and accountability in corporate reporting and disclosures, web3 companies face heightened scrutiny and compliance requirements. From financial reporting obligations to disclosure requirements for material events and cybersecurity incidents, web3 enterprises must navigate a labyrinth of regulatory mandates while upholding transparency and trustworthiness in their operations.

In addition to domestic regulations, Chapter 34 sheds light on the challenges posed by international security regulations for web3 companies with a global footprint. In an interconnected world where digital operations transcend national borders, web3 enterprises grapple with divergent regulatory regimes, geopolitical tensions, and cross-border data transfer restrictions. Navigating the complexities of international security regulations demands a nuanced understanding of jurisdictional differences, data sovereignty concerns, and the evolving landscape of global cybersecurity norms.

Finally, the chapter underscores the dynamic nature of regulatory changes and their impact on web3 companies, highlighting the need for agility, adaptability, and proactive compliance strategies. Regulatory frameworks governing privacy, data security, information disclosure, and international security are subject to frequent revisions, amendments, and updates in response to evolving threats, technological advancements, and geopolitical developments. Web3 companies must remain vigilant, anticipate regulatory changes, and proactively adjust their compliance efforts to ensure continued adherence to regulatory requirements and mitigate compliance risks.

Privacy Regulations and Data Security

Regulations on privacy and data security wield considerable influence over the digital security landscape of web3 companies, both positively and adversely. On the positive side, stringent privacy regulations such as the GDPR and CCPA compel web3 companies to prioritize data protection and implement robust cybersecurity measures. By mandating data minimization, encryption, and user consent mechanisms, these regulations enhance user privacy and bolster the overall security posture of web3 enterprises.[597] Compliance with privacy regulations fosters trust among users, enhances brand reputation, and mitigates the risk of data breaches and regulatory penalties, thereby fortifying the digital security of web3 companies.

Furthermore, privacy regulations empower individuals with greater control over their personal data, promoting transparency and accountability in data processing practices. Web3 companies are compelled to adopt privacy-by-design principles, conduct privacy impact assessments, and provide clear and concise privacy notices to users. By empowering individuals to exercise their privacy rights, privacy regulations foster a culture of data stewardship and responsible data management within web3 companies, thereby strengthening digital security and engendering trust among users.

[597] Ma, W. and Huang, K., 2022. *Blockchain and Web3: Building the cryptocurrency, privacy, and security foundations of the metaverse*. John Wiley & Sons.

However, despite these positive impacts, privacy regulations also pose challenges and unintended consequences that can adversely affect the digital security of web3 companies. Compliance with stringent regulatory requirements often entails significant costs, resource allocation, and operational complexities for web3 enterprises.[598] The implementation of privacy-enhancing technologies, data protection measures, and compliance frameworks necessitates substantial investments in infrastructure, training, and governance, diverting resources away from other critical security initiatives.

The fragmented nature of privacy regulations across different jurisdictions complicates compliance efforts for web3 companies operating in multiple geographic regions. Divergent legal frameworks, conflicting requirements, and jurisdictional ambiguities create compliance challenges and regulatory uncertainties, increasing the compliance burden and operational risks for web3 enterprises.[599] The extraterritorial reach of privacy regulations, such as the GDPR's applicability to non-EU companies processing EU residents' data, further exacerbates compliance challenges and regulatory complexity for web3 companies with a global footprint.

Additionally, privacy regulations can inadvertently hinder cybersecurity collaboration and threat intelligence sharing among web3 companies and industry stakeholders. The stringent data protection requirements and legal constraints imposed by privacy regulations may deter organizations from sharing threat information, vulnerability insights, and best practices for fear of violating privacy laws or exposing

---

598 *Id.*

599 Bentazar, W., 2024. Decentralized Autonomous Organizations in Web3: Navigating the Legal Landscape of Privacy and Data Protection.

sensitive data to regulatory scrutiny.[600] This lack of collaboration and information sharing undermines collective cybersecurity efforts, weakens cyber resilience, and leaves web3 companies more vulnerable to sophisticated cyber threats and attacks.

Furthermore, privacy regulations may incentivize the adoption of encryption and anonymization techniques to protect sensitive data, which can inadvertently impede cybersecurity monitoring and threat detection capabilities. Encrypted data poses challenges for traditional security tools and techniques, hindering the detection of malicious activities, insider threats, and data exfiltration attempts within web3 environments.[601] As a result, web3 companies may face heightened cybersecurity risks and operational challenges in detecting and mitigating security incidents, despite their compliance with privacy regulations aimed at protecting user privacy and data security.

Regulations Related to Turning over Information for Warrants

Regulations pertaining to turning over information in warrants wield significant influence over the digital security landscape of companies, impacting their operations, compliance efforts, and overall cybersecurity posture. On one hand, compliance with lawful requests for information under warrants is essential for upholding the rule of law, facilitating criminal investigations, and ensuring public safety.[602] By cooperating with law

---

[600] *Id.*
[601] Bambacht, J. and Pouwelse, J., 2022. Web3: A decentralized societal infrastructure for identity, trust, money, and data. *arXiv preprint arXiv:2203.00398.*
[602] Garon, J.M., 2022. Legal implications of a ubiquitous metaverse and a Web3 future. *Marq. L. Rev.*, *106*, p.163.

enforcement agencies and judicial authorities, companies play a crucial role in combating cybercrime, terrorist activities, and other illicit activities that pose threats to digital security.

However, the process of turning over information in warrants presents a myriad of challenges and potential risks for companies, particularly in the realm of digital security. First and foremost, compliance with warrant requests may entail disclosing sensitive or confidential information that could compromise the security and privacy of individuals, customers, or business operations.[603] In cases where companies store vast amounts of personal data or proprietary information, the disclosure of such data pursuant to a warrant could expose them to reputational damage, legal liabilities, and regulatory penalties.

The disclosure of information under warrants may disrupt ongoing investigations, compromise evidentiary integrity, and undermine cybersecurity efforts aimed at protecting sensitive data from unauthorized access or disclosure.[604] Companies must balance the imperative of law enforcement cooperation with their obligations to safeguard user privacy, data security, and due process rights. Failure to strike this balance effectively could result in adverse consequences for both the company and its stakeholders, including erosion of trust, loss of business opportunities, and legal ramifications.

Furthermore, compliance with warrant requests may necessitate the development of robust processes, protocols,

---

[603] Nwogugu, M.C., 2022. ESG, Inequality And International Political Economy Problems Inherent In Regulation Of DeFi, Web5/Web3, CryptoCurrencies And Fractional-NFTs. *Web5/Web3, CryptoCurrencies And Fractional-NFTs (Revised 2022).*
[604] *Id.*

and safeguards to ensure the lawful and secure handling of information subject to disclosure.[605] Companies must establish clear procedures for validating warrant requests, verifying the authenticity of requests, and ensuring compliance with legal requirements and procedural safeguards. Additionally, companies may need to invest in encryption technologies, access controls, and data protection measures to safeguard information subject to disclosure and mitigate the risk of unauthorized access or misuse.

Additionally, the extraterritorial reach of warrant requests and cross-border data transfer requirements pose significant challenges for companies operating in global markets. Jurisdictional differences, conflicting legal frameworks, and data sovereignty concerns complicate the process of complying with warrant requests while adhering to international legal standards and regulatory requirements. Companies must navigate a complex legal landscape, reconcile conflicting obligations, and mitigate legal and operational risks associated with cross-border data transfers and jurisdictional disputes.[606] Failure to navigate these challenges effectively could result in regulatory scrutiny, legal disputes, and reputational damage for the company.

Regulations for Publicly Disclosed Information

Regulations governing the public disclosure of information pose a complex security risk for web3 companies, as they navigate the delicate balance between transparency, accountability, and cybersecurity. On one hand, public disclosure regulations promote transparency and trust by

---

[605] Kushnir, A., Rennie, E., Holcombe-James, I., Kushnir, A., Webster, T., Morgan, B.A. and DOI, R., 2022. Developments in Web3 for the Creative Industries.
[606] *Id.*

providing stakeholders with access to timely and accurate information about the company's operations, financial performance, and governance practices. By fostering transparency, public disclosure regulations enhance market integrity, investor confidence, and regulatory compliance, thereby contributing to a robust and trustworthy digital ecosystem.

However, the public disclosure of sensitive or proprietary information under regulatory mandates poses significant security risks for web3 companies, exposing them to a range of threats, including intellectual property theft, competitive intelligence gathering, and reputational harm. Disclosure requirements may compel companies to divulge confidential business strategies, trade secrets, and proprietary technologies, thereby exposing them to the risk of exploitation by competitors, malicious actors, or hostile entities.[607] The public disclosure of security vulnerabilities, data breaches, or regulatory non-compliance incidents could tarnish the company's reputation, erode stakeholder trust, and undermine its competitive position in the market.

To mitigate the security risks associated with public disclosure regulations, web3 companies must adopt a proactive and strategic approach to managing information disclosure and safeguarding sensitive data. First and foremost, companies should conduct comprehensive risk assessments to identify and prioritize the security risks associated with public disclosure obligations. By understanding the potential impact of disclosure requirements on their operations, intellectual property, and competitive advantage, companies can develop tailored risk

---

[607] Bambacht, J. and Pouwelse, J., 2022. Web3: A decentralized societal infrastructure for identity, trust, money, and data. *arXiv preprint arXiv:2203.00398*

mitigation strategies and security controls to protect sensitive information from unauthorized access, disclosure, or exploitation.

Web3 companies should implement robust information governance frameworks and compliance programs to ensure the lawful and secure handling of information subject to public disclosure. This entails establishing clear policies, procedures, and controls for managing disclosure obligations, vetting information for accuracy and completeness, and safeguarding sensitive data from unauthorized access or disclosure.[608] Additionally, companies should invest in employee training and awareness programs to educate staff about their responsibilities under public disclosure regulations and reinforce best practices for safeguarding sensitive information.

Furthermore, web3 companies should leverage technology solutions, such as encryption, access controls, and data loss prevention tools, to enhance the security of information subject to public disclosure. Encryption technologies can help protect sensitive data in transit and at rest, while access controls and authentication mechanisms can restrict access to authorized personnel only. Additionally, data loss prevention tools can help detect and prevent unauthorized disclosures of sensitive information, providing an added layer of security against insider threats and external attacks.

While public disclosure regulations play a vital role in promoting transparency and accountability in the digital environment, they also pose significant security risks for web3 companies. By adopting a proactive and strategic approach to managing information disclosure,

---

[608] *Id.*

implementing robust governance frameworks and compliance programs, and leveraging technology solutions to enhance data security, web3 companies can mitigate the security risks associated with public disclosure regulations and safeguard sensitive information from unauthorized access or exploitation.

Regulations for Internal Security

Regulations concerning internal security and the privacy of data captured by cameras hold paramount importance for web3 companies operating in the global market. As digital ecosystems become increasingly interconnected and reliant on surveillance technologies, web3 enterprises must navigate a complex regulatory landscape governing the collection, storage, and use of video data to ensure compliance, safeguard privacy rights, and mitigate security risks.

First and foremost, regulations related to internal security and data privacy serve as essential safeguards to protect individuals' privacy rights and personal data from unauthorized access, misuse, or exploitation. With the proliferation of surveillance cameras in public and private spaces, there is a pressing need for regulatory frameworks that delineate clear guidelines and standards for the ethical and responsible use of video data.[609] Compliance with these regulations is essential for maintaining trust among customers, employees, and stakeholders, thereby safeguarding the reputation and credibility of web3 companies in the global market.

[609] Bambacht, Joost, and Johan Pouwelse. "Web3: A decentralized societal infrastructure for identity, trust, money, and data." *arXiv preprint arXiv:2203.00398* (2022).

Regulations governing internal security and data privacy provide a legal framework for web3 companies to establish and enforce robust data protection measures, access controls, and encryption protocols to safeguard video data from cyber threats, data breaches, and unauthorized disclosures. By adhering to regulatory requirements, web3 enterprises can demonstrate their commitment to data privacy and security, thereby enhancing customer confidence, mitigating regulatory risks, and avoiding potential legal liabilities associated with non-compliance.

Furthermore, compliance with internal security and data privacy regulations fosters a culture of accountability, transparency, and responsible data management within web3 companies. By implementing privacy-by-design principles and conducting privacy impact assessments, companies can proactively identify and mitigate privacy risks associated with the collection and processing of video data.[610] Additionally, transparency in data handling practices and clear communication with stakeholders about privacy policies and procedures help build trust and confidence in the company's commitment to protecting personal data.

In addition to ethical considerations, adherence to internal security and data privacy regulations is essential for web3 companies to maintain a competitive edge in the global market. As data protection regulations become increasingly stringent and enforcement actions more robust, companies that prioritize privacy and security are better positioned to differentiate themselves from competitors, attract customers, and foster long-term relationships based on trust and transparency.

---

[610] Huang, K., Budorin, D., Tan, L.J., Ma, W. and Zhang, Z.W., A Comprehensive Guide for Web3 Security..

In conclusion, regulations concerning internal security and data privacy are indispensable for web3 companies operating in the global market. By adhering to these regulations, companies can protect individuals' privacy rights, mitigate security risks, and demonstrate their commitment to ethical data practices. Moreover, compliance with internal security and data privacy regulations enhances customer trust, fosters transparency, and strengthens the company's competitive position in an increasingly data-driven and interconnected world.

The Ability for Regulations to Change on Short Notice

The regulatory landscape, characterized by jurisdictions where laws can swiftly evolve, presents both challenges and benefits for web3 businesses concerning digital security. One of the primary challenges stems from the inherent complexity and uncertainty associated with navigating rapidly changing legal frameworks. Web3 enterprises operating across multiple jurisdictions must continuously monitor legislative developments, anticipate regulatory changes, and adapt their compliance strategies accordingly. This dynamic environment poses challenges in terms of resource allocation, operational agility, and compliance readiness, as companies grapple with the need to interpret and implement evolving regulations while maintaining robust security measures.

The rapid pace of regulatory change introduces a degree of unpredictability and regulatory risk for web3 businesses. Legal uncertainties arising from frequent legislative updates, amendments, and revisions can create compliance gaps, regulatory ambiguities, and enforcement challenges for companies. The lack of clarity regarding compliance requirements and regulatory expectations may lead to inadvertent non-compliance, exposure to regulatory penalties, and legal disputes, thereby undermining the digital security and operational stability of web3 enterprises.

Among the challenges, there are also opportunities for web3 businesses to leverage the agility and adaptability inherent in their digital infrastructure to address regulatory changes proactively and enhance digital security. By adopting a proactive approach to compliance, web3 enterprises can leverage technology solutions, such as automated compliance monitoring tools, data analytics, and regulatory intelligence platforms, to streamline compliance processes, identify regulatory risks, and implement timely remediation measures.

Furthermore, the dynamic regulatory landscape offers opportunities for innovation and market differentiation, as web3 businesses develop innovative solutions to address emerging regulatory challenges and enhance digital security. Companies that embrace regulatory compliance as a strategic imperative and invest in proactive compliance initiatives can gain a competitive advantage by demonstrating their commitment to data protection, privacy, and security to customers, partners, and investors.

Additionally, the rapid pace of regulatory change necessitates enhanced collaboration and information-sharing among industry stakeholders, regulatory authorities, and policymakers to foster regulatory alignment, consistency, and interoperability. By engaging in constructive dialogue with regulators, participating in industry forums, and sharing best practices, web3 businesses can contribute to the development of coherent regulatory frameworks that promote digital security while facilitating innovation and growth in the digital economy.

While navigating jurisdictions where laws can change quickly poses challenges for web3 businesses concerning digital security, it also presents opportunities for innovation, agility, and market differentiation. By adopting a proactive approach to compliance, leveraging technology solutions, and fostering collaboration with regulatory authorities and industry peers, web3 enterprises can navigate the complexities of the regulatory landscape, enhance digital security, and position themselves for success in a rapidly evolving digital environment.

Conclusion

Chapter 34 has provided a comprehensive examination of the regulatory landscape surrounding internal security and the privacy of data captured by cameras, shedding light on the complexities and challenges faced by web3 companies operating in the global market. We have explored the critical importance of compliance with regulations governing data privacy, surveillance, and information disclosure, as well as the inherent risks and opportunities associated with navigating rapidly changing legal frameworks.

Throughout this chapter, we have underscored the significance of striking a delicate balance between transparency, accountability, and digital security in the collection, storage, and use of video data. By adhering to regulatory requirements, implementing robust security measures, and fostering a culture of privacy and data protection, web3 companies can build trust, mitigate risks, and enhance their competitive position in the global market.

Moreover, we have highlighted the challenges posed by regulatory uncertainties, jurisdictional differences, and the rapid pace of legislative change, while also acknowledging the opportunities for innovation, collaboration, and market differentiation that arise in dynamic regulatory environments.

As web3 companies continue to navigate the complexities of the regulatory landscape, it is imperative that they remain vigilant, proactive, and adaptable in their compliance efforts. By staying abreast of regulatory developments, leveraging technology solutions, and engaging in constructive dialogue with regulators and industry peers, web3 enterprises can navigate the intricacies of internal security and data privacy regulations while safeguarding the privacy rights and digital security of their stakeholders.

In essence, Chapter 34 serves as a call to action for web3 companies to embrace regulatory compliance as a strategic imperative, prioritize digital security and data privacy in their operations, and foster a culture of ethical data management and responsible surveillance practices. By doing so, web3 enterprises can not only mitigate regulatory risks and legal liabilities but also foster trust, build resilience, and unlock new opportunities for growth and innovation in the global digital economy.

Chapter 35: Cross-Border Cyber Crime

"One of the main cyber-risks is to think they don't exist. The other is to try to treat all potential risks. Fix the basics, protect first what matters for your business and be ready to react properly to pertinent threats. Think data, but also business services integrity, awareness, customer experience, compliance, and reputation."[611]

Chapter 35 diagnosis the intricate world of cross-border cybercrime, where the digital landscape is rife with threats ranging from data breaches and ransomware attacks to supply chain infiltrations and state-sponsored cyber espionage. As web3 companies navigate the complexities of the global market, they encounter an array of cyber threats that transcend geographical boundaries and challenge traditional notions of cybersecurity. In this chapter, we embark on a comprehensive exploration of the various tactics, techniques, and procedures employed by cybercriminals to exploit vulnerabilities, compromise digital assets, and disrupt business operations.

Data breaches and exfiltration represent one of the most prevalent and damaging forms of cybercrime, wherein malicious actors gain unauthorized access to sensitive data and subsequently exfiltrate it for nefarious purposes. Whether motivated by financial gain, espionage, or ideological motives, data breaches pose significant risks to web3 companies, including reputational damage, legal liabilities, and regulatory penalties. By understanding the mechanisms and techniques used by cybercriminals to infiltrate networks, steal data, and evade detection, web3

[611] (2024). "Cyber Crime Quotes, Stephene Nappo." Good Reads at https://www.goodreads.com/quotes/tag/cybercrime.

enterprises can enhance their cybersecurity posture and mitigate the risk of data breaches.

Ransomware has emerged as a pervasive and disruptive threat to web3 companies, leveraging encryption techniques to lock down critical systems and extort ransom payments from victims. The proliferation of ransomware-as-a-service (RaaS) models and anonymous cryptocurrency payments has facilitated the widespread adoption of ransomware by cybercriminals, leading to unprecedented levels of business disruption and financial loss. By examining the anatomy of ransomware attacks, the impact on business operations, and effective mitigation strategies, web3 companies can fortify their defenses and mitigate the risk of falling victim to ransomware extortion.

Supply chain attacks represent another insidious form of cybercrime, wherein attackers target third-party vendors or service providers to gain unauthorized access to target organizations' networks and systems. By exploiting vulnerabilities in the supply chain ecosystem, cybercriminals can infiltrate trusted networks, bypass perimeter defenses, and compromise sensitive data or critical infrastructure. Web3 companies must adopt a holistic approach to supply chain security, including rigorous vendor vetting processes, ongoing risk assessments, and robust security controls to mitigate the risk of supply chain attacks.

Social engineering tactics, such as phishing, spear phishing, and pretexting, pose significant challenges for web3 companies by exploiting human vulnerabilities to gain unauthorized access to sensitive information or systems. By impersonating trusted entities, manipulating emotions, and leveraging psychological tactics, cybercriminals can deceive employees, customers, or partners into disclosing confidential information or performing unauthorized actions. Web3 enterprises must invest in employee training, awareness programs, and technical controls to detect and mitigate the risk of social engineering attacks and safeguard against human error.

Intellectual property theft represents a pervasive and damaging form of cybercrime, wherein attackers target proprietary information, trade secrets, or research and development data for theft or exploitation. Whether perpetrated by nation-state actors, competitors, or insiders, intellectual property theft poses significant risks to web3 companies, including loss of competitive advantage, reputational damage, and financial harm. By implementing robust data protection measures, access controls, and intellectual property rights management strategies, web3 enterprises can mitigate the risk of intellectual property theft and safeguard their valuable assets.

Payment scams, including business email compromise (BEC), invoice fraud, and payment redirection schemes, represent a significant threat to web3 companies by exploiting vulnerabilities in financial processes and payment systems. By impersonating legitimate entities, manipulating communications, and deceiving employees, cybercriminals can orchestrate sophisticated payment scams to defraud organizations of funds or sensitive information. Web3 companies must implement robust financial controls, verification procedures, and employee training to detect and mitigate the risk of payment scams and prevent financial losses.

Finally, cyber espionage, often conducted by nation-state actors or advanced persistent threat (APT) groups, represents a formidable threat to web3 companies by targeting sensitive information, intellectual property, or critical infrastructure for espionage or sabotage. State-sponsored cyber espionage campaigns leverage sophisticated tactics, techniques, and procedures to infiltrate target networks, gather intelligence, and achieve strategic objectives. By understanding the motivations and tactics employed by cyber espionage actors, web3 enterprises can enhance their threat intelligence capabilities, detect intrusions, and mitigate the risk of cyber espionage attacks.

Chapter 35 provides a comprehensive overview of cross-border cybercrime, examining the various tactics, techniques, and procedures used by cybercriminals to exploit vulnerabilities, compromise digital assets, and disrupt business operations. By understanding the evolving threat landscape and implementing effective cybersecurity measures, web3 companies can mitigate the risk of cybercrime and safeguard their digital assets, intellectual property, and reputation in an increasingly interconnected and hostile digital environment.

Data Breaches and Exfiltration

Data breaches and exfiltration present significant challenges and consequences for web3 companies, affecting their operations, reputation, and bottom line. When sensitive data is compromised and extracted from a company's systems without authorization, it can lead to severe financial losses, legal liabilities, and damage to brand trust.[612] For web3 companies that often handle large volumes of user data and proprietary information, the impact of a data breach can be particularly devastating, as it

---

[612] Kshetri, N., 2023. Privacy violations, security breaches and other threats of Web3 and the metaverse.

may result in regulatory fines, customer churn, and loss of competitive advantage.

One of the primary concerns arising from data breaches and exfiltration is the potential exposure of sensitive customer data, including personal identifiable information (PII), financial records, and confidential communications. Unauthorized access to such data can result in identity theft, fraud, and other forms of cybercrime, leading to financial harm and reputational damage for both the company and its customers. Additionally, data breaches may violate data protection regulations, such as the GDPR or CCPA, exposing web3 companies to regulatory fines, legal penalties, and lawsuits.

To prevent damage from data breaches and exfiltration, web3 companies must adopt a multi-layered approach to cybersecurity, incorporating both technical controls and organizational practices to mitigate risks and protect sensitive data. One of the key methods used to prevent data breaches is the implementation of robust access controls and encryption mechanisms to restrict unauthorized access to sensitive information.[613] By encrypting data at rest and in transit, companies can mitigate the risk of data theft and unauthorized disclosure, even in the event of a breach.

Furthermore, web3 companies can leverage advanced threat detection and monitoring tools to detect and respond to suspicious activities and potential data breaches in real-time. Intrusion detection systems (IDS), security information and event management (SIEM) platforms, and behavior analytics solutions can help companies identify anomalous behavior, unauthorized access attempts, and

---

[613] *Id.*

data exfiltration activities, enabling timely intervention and mitigation of security incidents.

Additionally, employee training and awareness programs play a crucial role in preventing data breaches by educating staff about cybersecurity best practices, recognizing phishing attempts, and understanding the importance of data protection and confidentiality. By fostering a culture of security awareness and accountability, web3 companies can empower employees to identify and report potential security threats, reducing the likelihood of successful data breaches resulting from human error or negligence.

Moreover, regular security audits, vulnerability assessments, and penetration testing exercises can help web3 companies identify and address security weaknesses and vulnerabilities in their systems and infrastructure proactively.[614] By conducting comprehensive security assessments and implementing remediation measures, companies can strengthen their defenses, reduce attack surfaces, and minimize the risk of data breaches and exfiltration.

Data breaches and exfiltration pose significant risks and consequences for web3 companies, including financial losses, reputational damage, and regulatory penalties. To mitigate these risks, web3 companies must adopt a proactive approach to cybersecurity, incorporating robust technical controls, organizational practices, and employee training to prevent unauthorized access to sensitive data, detect security incidents, and respond effectively to potential breaches. By implementing these preventive

---

[614] Parisi, C. and Budorin, D., 2023. Web3 Security Analytics. In *A Comprehensive Guide for Web3 Security: From Technology, Economic and Legal Aspects* (pp. 159-176). Cham: Springer Nature Switzerland.

measures and maintaining a strong security posture, web3 companies can safeguard their digital assets, protect customer trust, and maintain compliance with data protection regulations.

Ransomware and Business Disruption

Ransomware has emerged as one of the most pervasive and disruptive forms of cybercrime, posing significant challenges to web3 companies and causing widespread business disruptions. This malicious software encrypts files and systems, rendering them inaccessible until a ransom is paid to the attackers, often in cryptocurrency. The impact of ransomware attacks on web3 companies can be severe, leading to financial losses, operational downtime, reputational damage, and legal liabilities.

One of the primary ways ransomware causes business disruptions is by encrypting critical systems and data, thereby preventing employees from accessing essential files, applications, and services necessary for day-to-day operations. This can result in significant productivity losses, as employees are unable to perform their tasks or access vital information needed to fulfill customer orders, process transactions, or maintain business continuity.[615] Ransomware attacks may disrupt supply chain operations, disrupt customer service, and damage business relationships, leading to long-term consequences for the company's reputation and competitiveness.

---

[615] Karapapas, C., Pittaras, I., Fotiou, N. and Polyzos, G.C., 2020, May. Ransomware as a service using smart contracts and IPFS. In *2020 IEEE International Conference on Blockchain and Cryptocurrency (ICBC)* (pp. 1-5). IEEE.

Furthermore, ransomware attacks often result in operational downtime as companies scramble to contain the breach, recover encrypted data, and restore systems to a functional state. This downtime can have a cascading effect on business operations, leading to missed deadlines, delayed projects, and revenue loss.[616] In some cases, ransomware attacks may disrupt critical infrastructure, such as manufacturing facilities, healthcare systems, or transportation networks, posing risks to public safety and national security.

To prevent ransomware attacks and mitigate their impact, web3 companies must adopt a proactive and multi-layered approach to cybersecurity. One of the primary methods used to prevent ransomware attacks is the implementation of robust endpoint protection solutions, such as antivirus software, intrusion detection systems, and endpoint detection and response (EDR) platforms. These solutions help detect and block ransomware threats at the endpoint level, preventing malware from executing and encrypting files.

Web3 companies can leverage email security solutions, such as spam filters, email encryption, and email authentication protocols, to protect against phishing attacks and malicious email attachments used to distribute ransomware. By implementing email security best practices and educating employees about the risks of phishing, companies can reduce the likelihood of falling victim to ransomware attacks initiated through email-based vectors.

---

[616] Winter, P., Lorimer, A.H., Snyder, P. and Livshits, B., 2021. Security, Privacy, and Decentralization in Web3. *arXiv preprint arXiv:2109.06836.*

Additionally, regular data backups and disaster recovery planning are essential components of a ransomware prevention strategy, allowing companies to restore encrypted data and systems from backup copies in the event of a ransomware attack. By maintaining offline backups stored in secure locations and testing backup and recovery procedures regularly, companies can minimize the impact of ransomware attacks and expedite the restoration process, reducing downtime and business disruptions.

Furthermore, network segmentation, access controls, and least privilege principles can help limit the spread of ransomware within corporate networks and prevent attackers from moving laterally to infect additional systems. By segmenting networks into smaller, isolated zones and restricting access to sensitive resources based on user roles and permissions, companies can contain ransomware infections and mitigate the risk of widespread data encryption and disruption.

Moreover, employee training and awareness programs play a crucial role in preventing ransomware attacks by educating staff about cybersecurity best practices, recognizing phishing attempts, and understanding the consequences of ransomware infections. By fostering a culture of security awareness and accountability, companies can empower employees to identify and report potential security threats, reducing the likelihood of successful ransomware attacks resulting from human error or negligence.

In conclusion, ransomware represents a significant threat to web3 companies, causing business disruptions, financial losses, and reputational damage. To prevent ransomware attacks and mitigate their impact, web3 companies must adopt a proactive and multi-layered approach to cybersecurity, incorporating robust endpoint protection, email security, data backups, network segmentation, and employee training. By implementing these preventive measures and maintaining a strong security posture, web3 companies can safeguard their digital assets, protect customer trust, and minimize the risk of ransomware-related disruptions.

Supply Chain Attacks

Supply chain attacks represent a sophisticated form of cyber threat where malicious actors target vulnerabilities in a company's supply chain ecosystem to compromise its systems, infrastructure, or data. These attacks leverage the interconnected nature of supply chains, exploiting trust relationships between vendors, suppliers, and partners to infiltrate target organizations and achieve their objectives. Supply chain attacks can take various forms, including the insertion of malicious hardware or software into supply chain components, tampering with software updates or patches, or compromising third-party service providers to gain unauthorized access to target networks.

One of the primary challenges posed by supply chain attacks is the difficulty of detecting and mitigating threats that originate outside the organization's immediate control. Unlike traditional cyber attacks that target internal systems and networks, supply chain attacks often exploit trust relationships and dependencies across multiple entities, making them harder to detect and defend against.[617] Supply chain attacks can have far-reaching consequences,

---

617 Cui, Y., Singh, P.J. and Li, H., 2024. Improving Supply Chain Resilience Through a Blockchain-Based System Architecture. In *Digitization in supply chain management: trends, challenges and solutions* (pp. 75-101).

impacting not only the targeted organization but also its customers, partners, and stakeholders.

To prevent supply chain attacks, companies must adopt a proactive and multi-faceted approach to supply chain security, incorporating robust risk management practices, vendor management protocols, and security controls to mitigate vulnerabilities and safeguard against potential threats. One of the key strategies companies can employ to prevent supply chain attacks is to conduct thorough risk assessments and due diligence when selecting and onboarding third-party vendors, suppliers, and service providers. By evaluating vendors' security posture, practices, and compliance with industry standards, companies can identify potential risks and take proactive measures to mitigate them before onboarding them into their supply chain ecosystem.

Furthermore, companies can implement supply chain security controls and best practices to enhance visibility, transparency, and accountability across their supply chain operations. This includes implementing supply chain risk management frameworks, such as the NIST Cybersecurity Framework or ISO 27001, to identify, assess, and manage risks throughout the supply chain lifecycle.[618] Additionally, companies can establish contractual agreements, service-level agreements (SLAs), and security requirements with vendors and suppliers to ensure compliance with security standards, data protection regulations, and incident response procedures.

---

[618] Sable, N.P., Sonkamble, R., Rathod, V.U., Shirke, S., Deshmukh, J.Y. and Chavan, G.T., 2023. Web3 Chain Authentication and Authorization Security Standard (CAA). *International Journal on Recent and Innovation Trends in Computing and Communication*, *11*(5), pp.70-76.

Companies can enhance supply chain resilience by diversifying their supplier base, reducing dependencies on single-source suppliers, and implementing redundancy and contingency plans to mitigate the impact of supply chain disruptions.[619] By diversifying suppliers and establishing alternative sourcing options, companies can reduce the risk of supply chain attacks and minimize the potential impact of disruptions caused by cyber threats or other external factors.

Additionally, continuous monitoring, threat intelligence, and incident response capabilities are essential components of a comprehensive supply chain security strategy. By implementing robust monitoring tools, security analytics platforms, and threat detection mechanisms, companies can proactively detect and respond to suspicious activities, anomalies, and indicators of compromise within their supply chain ecosystem. Furthermore, companies can establish incident response plans, communication protocols, and coordination mechanisms to facilitate timely response and recovery efforts in the event of a supply chain attack or security breach.

Supply chain attacks represent a significant threat to companies operating in the digital ecosystem, requiring proactive measures and strategic investments in supply chain security to mitigate risks and safeguard against potential threats. By adopting a comprehensive approach to supply chain security, including risk assessments, vendor management, security controls, resilience planning, and incident response capabilities, companies can enhance their ability to detect, prevent, and respond to supply chain attacks, thereby safeguarding their operations, assets, and

---

[619] *Id.*

reputation in an increasingly interconnected and volatile business environment.

Cryptocurrency Theft and Fraud

Cryptocurrency theft and fraud have become significant challenges in the web3 space, with malicious actors exploiting vulnerabilities in blockchain networks, cryptocurrency exchanges, and digital wallets to steal funds and defraud users.[620] These attacks take various forms, including hacking attacks targeting exchanges and wallet providers, phishing scams, Ponzi schemes, and fraudulent initial coin offerings (ICOs). The decentralized and pseudonymous nature of cryptocurrencies presents unique challenges for combating theft and fraud, as transactions are irreversible and identities are often difficult to trace.

One of the primary methods used by malicious actors to steal cryptocurrency is through hacking attacks targeting cryptocurrency exchanges and digital wallet providers. By exploiting vulnerabilities in exchange platforms or compromising users' credentials through phishing or malware attacks, hackers can gain unauthorized access to users' accounts and transfer funds to external wallets controlled by the attackers.[621] These attacks have resulted in significant financial losses for both individual users and cryptocurrency exchanges, undermining trust and confidence in the security of digital assets.

Phishing scams represent another prevalent form of cryptocurrency theft and fraud, wherein attackers use social

---

[620] Wu, J., Lin, K., Lin, D., Zheng, Z., Huang, H. and Zheng, Z., 2023. Financial crimes in web3-empowered metaverse: Taxonomy, countermeasures, and opportunities. *IEEE Open Journal of the Computer Society*, *4*, pp.37-49.
[621] *Id.*

engineering tactics to deceive users into disclosing their private keys, seed phrases, or login credentials.[622] These phishing attacks often take the form of fraudulent emails, websites, or social media accounts impersonating legitimate cryptocurrency exchanges, wallets, or projects. By tricking users into providing sensitive information or clicking on malicious links, attackers can gain access to users' funds and compromise their cryptocurrency holdings.

To improve security and mitigate the risk of cryptocurrency theft and fraud, companies operating in the web3 space must implement robust security measures and best practices to protect users' assets and safeguard against potential threats. One of the key strategies companies can employ is to prioritize the security of their platforms, exchanges, and wallets by implementing industry-standard security protocols, such as multi-factor authentication (MFA), encryption, and cold storage solutions.

Furthermore, companies can enhance user awareness and education about cryptocurrency security risks and best practices through comprehensive training programs, educational materials, and user guides. By educating users about the importance of securing their private keys, using reputable exchanges and wallets, and avoiding suspicious links or emails, companies can empower users to take proactive measures to protect their cryptocurrency holdings and mitigate the risk of theft or fraud.

Companies can implement stringent Know Your Customer (KYC) and Anti-Money Laundering (AML) procedures to verify the identity of users and prevent fraudulent activities on their platforms.[623] By conducting thorough identity

---

[622] *Id.*

[623] Bambacht, J. and Pouwelse, J., 2022. Web3: A decentralized

verification checks, monitoring transactions for suspicious activity, and complying with regulatory requirements, companies can reduce the risk of money laundering, fraud, and illicit activities involving cryptocurrencies.

Additionally, companies can collaborate with industry partners, cybersecurity firms, and law enforcement agencies to share threat intelligence, coordinate incident response efforts, and combat cryptocurrency theft and fraud collectively. By establishing partnerships and information-sharing networks, companies can enhance their ability to detect and mitigate emerging threats, identify malicious actors, and disrupt criminal activities in the web3 space.

Furthermore, companies can leverage blockchain analytics and forensic tools to trace stolen funds, track illicit transactions, and identify patterns of fraudulent activity on the blockchain. By analyzing transaction data, blockchain forensics experts can uncover evidence of theft, fraud, or money laundering, enabling law enforcement agencies to take appropriate action and recover stolen assets.

Cryptocurrency theft and fraud pose significant challenges for companies operating in the web3 space, requiring proactive measures and strategic investments in security to protect users' assets and maintain trust in the digital economy. By implementing robust security measures, enhancing user awareness and education, complying with regulatory requirements, collaborating with industry partners, and leveraging blockchain analytics and forensic tools, companies can improve their security posture and mitigate the risk of cryptocurrency theft and fraud, thereby safeguarding the integrity of the web3 ecosystem.

---

societal infrastructure for identity, trust, money, and data. *arXiv preprint arXiv:2203.00398.*

Social Engineering

Social engineering attacks represent a significant threat to modern businesses, leveraging psychological manipulation and deception to exploit human vulnerabilities and gain unauthorized access to sensitive information or systems. These attacks often target employees, customers, or partners through various communication channels, including email, phone calls, social media, or in-person interactions.[624] By impersonating trusted entities, creating a sense of urgency, or exploiting emotional triggers, social engineering attackers can deceive individuals into disclosing confidential information, clicking on malicious links, or performing unauthorized actions, thereby compromising the security of the organization.

To prevent social engineering attacks and mitigate their impact on modern businesses, companies must implement comprehensive security awareness programs and employee training initiatives to educate staff about the risks of social engineering and best practices for identifying and mitigating potential threats.[625] By raising awareness about common social engineering tactics, such as phishing, pretexting, and baiting, companies can empower employees to recognize suspicious behavior, question unexpected requests, and verify the legitimacy of communications before disclosing sensitive information or taking action.

Companies can implement technical controls and security measures to complement their employee training efforts and prevent social engineering attacks from succeeding. This includes deploying email security solutions, such as spam filters, antivirus software, and email authentication

[624] *Id.*
[625] *Id.*

protocols, to detect and block phishing attempts, malicious attachments, and suspicious links in email communications.[626] Additionally, companies can leverage endpoint protection solutions, intrusion detection systems, and security analytics platforms to monitor network traffic, detect anomalous behavior, and identify indicators of compromise associated with social engineering attacks. By combining employee education with robust technical controls, companies can enhance their resilience to social engineering threats and protect their assets, data, and reputation from malicious actors.[627]

IP Theft

Intellectual property (IP) theft in the web3 space poses significant challenges for companies, as digital assets and innovations are vulnerable to unauthorized duplication, distribution, or exploitation by malicious actors.[628] Intellectual property theft encompasses various forms, including copyright infringement, trademark counterfeiting, patent infringement, trade secret misappropriation, and plagiarism. In the web3 environment, where digital content, software, and innovations are easily replicable and distributable, the risk of IP theft is heightened, requiring companies to implement robust measures to protect their intellectual assets.

---

[626] Ivanov, N., Lou, J., Chen, T., Li, J. and Yan, Q., 2021, May. Targeting the weakest link: Social engineering attacks in ethereum smart contracts. In *Proceedings of the 2021 ACM asia conference on computer and communications security* (pp. 787-801).
[627] Wu, J., Lin, K., Lin, D., Zheng, Z., Huang, H. and Zheng, Z., 2023. Financial crimes in web3-empowered metaverse: Taxonomy, countermeasures, and opportunities. *IEEE Open Journal of the Computer Society*, *4*, pp.37-49.
[628] Beiyuan, Y.U., Shanyao, R.E.N. and Jianwei, L.I.U., 2023. Overview of blockchain assets theft attacks and defense technology. *Chinese Journal of Network & Information Security*, *9*(1).

Bad actors use several methods to engage in intellectual property theft in the web3 space, exploiting vulnerabilities in digital platforms, decentralized networks, and online marketplaces to illicitly access, copy, or monetize copyrighted content, patented inventions, or proprietary technologies.[629] These methods include unauthorized downloading and sharing of copyrighted materials, reverse engineering of software and hardware products to replicate functionality or design, and unauthorized distribution or sale of counterfeit goods bearing protected trademarks. Moreover, bad actors may engage in trade secret theft by infiltrating companies' networks or exploiting insider access to steal confidential business information, proprietary algorithms, or valuable customer data.

To avoid falling victim to intellectual property theft in the web3 space, companies must implement comprehensive strategies and safeguards to protect their intellectual assets and mitigate the risk of unauthorized exploitation or infringement. One key approach is to leverage technological solutions, such as digital rights management (DRM) systems, encryption techniques, and watermarking technologies, to secure digital content, software, and multimedia assets from unauthorized access, copying, or redistribution.[630] By encrypting files, embedding digital watermarks, or applying access controls, companies can deter unauthorized use and trace the origin of illicitly distributed intellectual property.

Furthermore, companies can establish legal protections and enforceable contracts to safeguard their intellectual property rights and deter potential infringers from engaging

---

[629] Kshetri, *supra*
[630] Voshmgir, S., 2020. *Token economy: How the Web3 reinvents the internet* (Vol. 2). Token Kitchen.

in unauthorized use or reproduction of protected assets. This includes registering copyrights, trademarks, and patents with relevant intellectual property offices, drafting licensing agreements, and implementing terms of use or service agreements that define permissible uses of digital assets and technologies. By asserting their intellectual property rights and enforcing legal remedies against infringers, companies can deter intellectual property theft and protect the value of their creations and innovations in the web3 ecosystem.

Companies can implement proactive monitoring and detection mechanisms to identify instances of intellectual property theft or infringement in the web3 space. This includes leveraging digital forensics tools, web scraping technologies, and online monitoring services to monitor digital channels, social media platforms, and online marketplaces for unauthorized use or distribution of copyrighted materials, patented inventions, or branded products.[631] By detecting and addressing instances of intellectual property theft in a timely manner, companies can mitigate the impact on their brand reputation, revenue streams, and competitive advantage.

Intellectual property theft in the web3 space poses significant risks and challenges for companies, requiring proactive measures and strategic investments in intellectual property protection and enforcement. By implementing technological solutions, legal protections, proactive monitoring, and enforcement measures, companies can safeguard their intellectual assets, deter potential infringers, and preserve the value of their creations and innovations in the digital economy.

---

[631] *Id.*

Payment Scams

Payment scams encompass a variety of fraudulent schemes aimed at deceiving businesses into making unauthorized payments or transferring funds to malicious actors posing as legitimate entities. These scams often exploit vulnerabilities in payment processes, financial controls, and communication channels to trick employees or vendors into sending money to fraudulent accounts or unauthorized recipients.[632] Common methods of payment scams include invoice fraud, business email compromise (BEC), CEO fraud, and vendor impersonation scams, where attackers use social engineering tactics, phishing emails, or spoofed communications to deceive victims into making payments under false pretenses.

Payment scams can have devastating consequences for businesses, resulting in financial losses, operational disruptions, reputational damage, and legal liabilities.[633] By exploiting trust relationships, impersonating trusted entities, or creating a sense of urgency, scammers can trick employees or vendors into transferring large sums of money to fraudulent accounts, often without the victim's knowledge or consent. Moreover, payment scams may lead to regulatory scrutiny, compliance violations, and breach of contractual obligations, further exacerbating the impact on affected businesses.[634]

---

632 Wu, J., Lin, K., Lin, D., Zheng, Z., Huang, H. and Zheng, Z., 2023. Financial crimes in web3-empowered metaverse: Taxonomy, countermeasures, and opportunities. *IEEE Open Journal of the Computer Society*, *4*, pp.37-49.
633 *Id.*
634 Bergquist Mcneil, L., 2022. Blockchains, smart contracts, and stablecoins as a global payment system: The rise of web 3.0.

To mitigate the risk of payment scams and protect against financial fraud, businesses can implement a range of preventive measures and security controls to detect and prevent unauthorized payments or fraudulent transactions. One key strategy is to establish robust authentication and authorization protocols for approving payments and fund transfers, including multi-factor authentication (MFA), dual authorization requirements, and segregation of duties to prevent unauthorized access or misuse of financial systems.[635] By implementing strong authentication mechanisms and access controls, businesses can reduce the risk of unauthorized payments and limit the impact of payment scams.

Furthermore, businesses can enhance employee awareness and training about the risks of payment scams and best practices for identifying and responding to suspicious requests or communications. By educating employees about common scam tactics, warning signs of fraudulent activity, and procedures for verifying payment requests, businesses can empower staff to exercise caution, skepticism, and due diligence when processing payments or responding to financial requests.[636] Regular training sessions, simulated phishing exercises, and communication campaigns can help reinforce security awareness and promote a culture of vigilance against payment scams.

Businesses can implement robust financial controls, segregation of duties, and approval workflows to prevent unauthorized payments and enhance oversight of financial transactions. This includes implementing transaction monitoring systems, anomaly detection mechanisms, and

---

[635] Winter, P., Lorimer, A.H., Snyder, P. and Livshits, B., 2021. Security, Privacy, and Decentralization in Web3. *arXiv preprint arXiv:2109.06836.*
[636] Wu et al, *supra.*

fraud detection algorithms to identify unusual patterns or suspicious activities indicative of payment scams. By implementing automated alerts, escalation procedures, and transaction reconciliation processes, businesses can detect and respond to potential payment scams in a timely manner, reducing the likelihood of financial losses and mitigating the impact on the organization.

Additionally, businesses can establish secure communication channels, verification protocols, and vendor validation processes to authenticate the identity of payment recipients and mitigate the risk of vendor impersonation scams or BEC attacks. By verifying the legitimacy of payment requests, validating vendor information, and cross-referencing banking details with trusted sources, businesses can reduce the risk of falling victim to payment scams and protect against unauthorized fund transfers or fraudulent transactions.[637] Overall, a multi-layered approach combining technical controls, employee training, financial oversight, and vendor validation can help businesses mitigate the risk of payment scams and protect against financial fraud in the digital age.

Cyber Espionage

Cyber espionage involves covert activities aimed at infiltrating, monitoring, and stealing sensitive information from target organizations, governments, or individuals for strategic, political, or economic gain. Commonly employed by nation-states, criminal organizations, or state-sponsored actors, cyber espionage utilizes various sophisticated techniques and tools to infiltrate target networks, exfiltrate sensitive data, and conduct intelligence gathering operations.[638] Some common methods used for cyber

---

[637] Kshetri, *supra*.

espionage include malware attacks, phishing campaigns, social engineering tactics, insider threats, supply chain attacks, and advanced persistent threats (APTs), each designed to exploit vulnerabilities and circumvent security defenses to achieve their objectives.

Malware attacks represent one of the most prevalent methods used for cyber espionage, wherein malicious software is deployed to infect target systems, steal sensitive information, or gain unauthorized access to network resources.[639] Malware variants such as spyware, keyloggers, remote access Trojans (RATs), and backdoors are commonly used by cyber espionage actors to covertly monitor and exfiltrate data from compromised systems. These malware payloads are often delivered through phishing emails, malicious websites, or software vulnerabilities, allowing attackers to establish persistent access and conduct surveillance activities within target networks.

Phishing campaigns are another common tactic employed in cyber espionage operations, wherein attackers use fraudulent emails, messages, or social media posts to deceive victims into disclosing sensitive information, credentials, or granting access to their systems.[640] By impersonating trusted entities, such as colleagues, vendors, or government agencies, attackers lure unsuspecting users into clicking on malicious links, downloading malware-infected attachments, or providing login credentials, thereby compromising their accounts or systems. Phishing

---

[638] Saracoglu, D., 2023. Metaverse and New Cybersecurity Threats. In *Metaverse: Technologies, Opportunities and Threats* (pp. 99-121). Singapore: Springer Nature Singapore.
[639] *Id.*
[640] Nath, K., 2023. Evolution of the Internet from Web 1. *Authorea Preprints.*

attacks can bypass traditional security controls and exploit human vulnerabilities to gain unauthorized access to target networks or steal sensitive data.

Social engineering tactics play a crucial role in cyber espionage operations, as attackers manipulate human behavior and psychological biases to exploit trust relationships, elicit sensitive information, or gain unauthorized access to target systems. Social engineering techniques such as pretexting, baiting, tailgating, and quid pro quo are commonly used to deceive individuals, employees, or insiders into disclosing confidential information, sharing access credentials, or performing unauthorized actions on behalf of the attackers.[641] By exploiting human weaknesses and exploiting trust, social engineering attacks can bypass technical controls and facilitate unauthorized access to sensitive data or systems.

Insider threats pose a significant risk to organizations engaged in cyber espionage, as trusted insiders with access to sensitive information or privileged accounts may betray their trust and collude with external adversaries to steal or compromise valuable assets. Insider threats can take various forms, including malicious insiders who intentionally sabotage systems or steal data for personal gain, as well as unwitting insiders who inadvertently expose sensitive information or fall victim to social engineering tactics.[642] To prevent insider threats, organizations must implement robust access controls, monitoring mechanisms, and employee awareness programs to detect and deter unauthorized access or misuse of privileged accounts.

---

[641] Badih, H., 2020. *A Blockchain and Deception Co-Design for Webcam Spyware Detection Protocol* (Doctoral dissertation, Oakland University).
[642] *Id.*

Supply chain attacks represent a sophisticated form of cyber espionage, wherein attackers target third-party vendors, suppliers, or service providers to gain unauthorized access to target networks or compromise the security of downstream organizations. By infiltrating trusted supply chain partners or exploiting vulnerabilities in software supply chains, attackers can implant malware, backdoors, or malicious code into legitimate software updates or components, allowing them to compromise target systems without direct infiltration.[643] Supply chain attacks pose significant challenges for organizations, as they bypass traditional perimeter defenses and exploit trust relationships between vendors and customers.

Advanced persistent threats (APTs) are sophisticated, long-term cyber espionage campaigns conducted by nation-states, intelligence agencies, or well-funded cybercriminal groups to infiltrate and compromise target networks for intelligence gathering or strategic purposes. APT actors employ a combination of stealthy tactics, custom malware, zero-day exploits, and advanced evasion techniques to evade detection, maintain persistent access, and exfiltrate sensitive data over extended periods.[644] APT campaigns often target high-value organizations, government agencies, or critical infrastructure sectors, leveraging a combination of technical expertise, social engineering tactics, and strategic intelligence to achieve their objectives.

To prevent cyber espionage and protect against unauthorized access, organizations must implement a multi-layered security strategy incorporating technical controls, employee awareness training, supply chain risk

---

[643] Vartanian, T.P., 2023. *The unhackable internet: How rebuilding cyberspace can create real security and prevent financial collapse*. Rowman & Littlefield.
[644] Wu, *supra*.

management, insider threat detection, and incident response capabilities. Technical controls such as firewalls, intrusion detection systems (IDS), endpoint security solutions, and data encryption can help detect and mitigate cyber espionage threats by preventing unauthorized access, detecting suspicious activities, and encrypting sensitive data. Employee awareness training programs should educate staff about the risks of cyber espionage, social engineering tactics, and best practices for identifying and reporting suspicious behavior.[645] Supply chain risk management practices can help organizations identify and mitigate vulnerabilities in their supply chain ecosystem, including third-party vendors, suppliers, or service providers. Insider threat detection mechanisms, such as user behavior analytics (UBA) and privileged access management (PAM), can help organizations monitor and mitigate the risk of insider threats by identifying anomalous behavior, restricting access to sensitive information, and enforcing least privilege principles. Finally, incident response capabilities, including incident detection, containment, eradication, and recovery, are essential for organizations to effectively respond to and mitigate the impact of cyber espionage incidents. By implementing these preventive measures and security controls, organizations can enhance their resilience to cyber espionage threats and protect against unauthorized access, data breaches, and intellectual property theft.

State Sponsored Attacks

State-sponsored attacks pose a significant danger to businesses, as they are often well-funded, sophisticated,

[645] Pattal, M.M.I., Li, Y. and Zeng, J., 2009, September. Web 3.0: A real personal web! More opportunities and more threats. In *2009 Third International Conference on Next Generation Mobile Applications, Services and Technologies* (pp. 125-128). IEEE.

and orchestrated by nation-states with geopolitical motives. These attacks are typically carried out by government-backed hackers or intelligence agencies seeking to steal sensitive information, disrupt operations, or sabotage critical infrastructure for political, economic, or strategic purposes. State-sponsored attackers leverage advanced cyber capabilities, zero-day exploits, and custom malware to infiltrate target organizations, evade detection, and achieve their objectives. The consequences of state-sponsored attacks can be severe, resulting in financial losses, reputational damage, regulatory scrutiny, and national security implications.

One method used to prevent state-sponsored attacks is to implement robust cybersecurity defenses and resilience measures to detect, deter, and mitigate advanced threats. This includes deploying next-generation firewalls, intrusion detection systems (IDS), and endpoint security solutions to monitor network traffic, detect anomalous behavior, and block malicious activities. By implementing a defense-in-depth strategy, organizations can create multiple layers of security controls to protect against sophisticated attack techniques used by state-sponsored adversaries.

Another approach to prevent state-sponsored attacks is to conduct regular security assessments, penetration testing, and vulnerability assessments to identify and remediate security weaknesses in the organization's infrastructure, applications, and systems. By proactively identifying and addressing vulnerabilities, organizations can reduce the attack surface and minimize the risk of exploitation by state-sponsored attackers. Additionally, organizations should prioritize patch management and software updates to ensure that systems are protected against known vulnerabilities and security flaws.

Furthermore, organizations can enhance employee awareness and training programs to educate staff about the risks of state-sponsored attacks, social engineering tactics, and best practices for cybersecurity hygiene.[646] By raising awareness about the tactics, techniques, and procedures (TTPs) used by state-sponsored adversaries, employees can become more vigilant and proactive in identifying and reporting suspicious activities. Regular training sessions, simulated phishing exercises, and incident response drills can help reinforce security awareness and readiness across the organization.

In addition to technical controls and employee training, organizations can strengthen their defenses against state-sponsored attacks by establishing threat intelligence sharing partnerships and collaboration networks with industry peers, government agencies, and cybersecurity organizations. By sharing threat intelligence, indicators of compromise (IOCs), and actionable insights about emerging threats, organizations can enhance their situational awareness, threat detection capabilities, and incident response readiness. Collaborative efforts to share threat intelligence can help organizations stay ahead of evolving threats and effectively defend against state-sponsored attacks.

Organizations should establish incident response plans, protocols, and procedures to facilitate timely detection, containment, and remediation of state-sponsored attacks. By defining roles and responsibilities, establishing communication channels, and conducting regular tabletop exercises, organizations can streamline incident response

---

[646] Torres, C.F., Willi, F. and Shinde, S., 2023. Is your wallet snitching on you? an analysis on the privacy implications of web3. In *32nd USENIX Security Symposium (USENIX Security 23)* (pp. 769-786).

efforts and minimize the impact of cyber incidents on business operations.[647] Additionally, organizations should develop relationships with law enforcement agencies, regulatory bodies, and cybersecurity experts to facilitate collaboration and support during incident response and investigation activities.

Furthermore, organizations can leverage encryption, data masking, and data loss prevention (DLP) solutions to protect sensitive information and intellectual property from state-sponsored espionage activities. By encrypting data at rest and in transit, implementing access controls, and monitoring data access and usage, organizations can mitigate the risk of data exfiltration and unauthorized access by state-sponsored attackers. Additionally, organizations should implement strong authentication mechanisms, such as multi-factor authentication (MFA) and biometric authentication, to prevent unauthorized access to sensitive systems and data.

State-sponsored attacks represent a serious threat to businesses, requiring proactive measures and strategic investments in cybersecurity to prevent, detect, and mitigate advanced threats. By implementing robust cybersecurity defenses, conducting regular security assessments, raising employee awareness, sharing threat intelligence, establishing incident response capabilities, and protecting sensitive information, organizations can enhance their resilience to state-sponsored attacks and safeguard against the potential impact on their operations, reputation, and national security interests.

---

[647] Garon, J.M., 2022. Legal implications of a ubiquitous metaverse and a Web3 future. *Marq. L. Rev.*, *106*, p.163.

Conclusion

In conclusion, Chapter 35 has provided a comprehensive overview of the multifaceted landscape of cross-border cybercrime, shedding light on the various methods employed by malicious actors to exploit vulnerabilities in the digital realm. From data breaches and exfiltration to ransomware attacks, supply chain compromises, and social engineering tactics, the chapter has underscored the diverse range of threats that businesses face in the interconnected digital ecosystem. Moreover, the discussion on intellectual property theft, payment scams, and cyber espionage has highlighted the pervasive nature of cyber threats and their implications for organizations operating in the web3 space.

Throughout the chapter, it has become evident that cybercrime poses significant risks and challenges for businesses, governments, and individuals alike. The evolving nature of cyber threats, coupled with the increasing sophistication of attackers, underscores the critical importance of robust cybersecurity measures and proactive risk management strategies. As organizations continue to digitize their operations and embrace emerging technologies, they must remain vigilant and proactive in safeguarding their assets, data, and systems against cyber threats.

Furthermore, the chapter has emphasized the need for collaboration and information sharing among stakeholders to combat cybercrime effectively. In an era where cyber threats transcend borders and jurisdictions, collaboration between government agencies, law enforcement authorities, private sector organizations, and cybersecurity experts is essential to effectively detect, deter, and respond to cyber threats. By fostering a culture of collaboration, knowledge sharing, and collective action, stakeholders can enhance their resilience to cyber threats and mitigate the impact of cybercrime on society as a whole.

The chapter has underscored the importance of proactive cybersecurity measures and risk mitigation strategies to protect against cyber threats. From implementing robust technical controls and employee training programs to enhancing incident response capabilities and establishing partnerships for threat intelligence sharing, organizations have a range of tools and approaches at their disposal to bolster their cyber defenses. By adopting a comprehensive and proactive approach to cybersecurity, organizations can better protect their assets, data, and reputation from cyber threats.

In summation, the fight against cross-border cybercrime requires a concerted effort and a multifaceted approach that combines technical, organizational, and collaborative strategies. As cyber threats continue to evolve and grow in sophistication, it is imperative for organizations to remain agile, adaptable, and proactive in their cybersecurity efforts. By staying informed, implementing best practices, and fostering collaboration across sectors and borders, stakeholders can effectively confront the challenges posed by cybercrime and build a more secure and resilient digital future.

Chapter 36: Sanctions and Surveillance

"The freedom to criticize judges and other public officials is necessary to a vibrant democracy. The problem comes when health criticism is replaced with more destructive intimidation and sanctions."[648]

In Chapter 36 we discuss sanctions and surveillance, exploring the multifaceted challenges that businesses encounter due to government actions. In an era of heightened geopolitical tensions and regulatory scrutiny, businesses must navigate a complex landscape of sanctions regimes and surveillance measures that can significantly impact their operations and growth prospects. This chapter sheds light on various aspects of sanctions and surveillance, offering insights into how businesses can proactively prepare and mitigate the risks associated with these regulatory frameworks.

One critical aspect addressed in this chapter is how businesses can strategize to avoid falling afoul of sanctions imposed by governments or international bodies. With sanctions regimes evolving and expanding, it is imperative for businesses to stay abreast of regulatory developments, conduct thorough due diligence on their business partners and transactions, and implement robust compliance programs to ensure adherence to sanctions laws and regulations. By implementing effective risk assessment frameworks and compliance controls, businesses can minimize the risk of inadvertently violating sanctions and facing severe penalties.

---

[648] (2024). "Sanctions Quotes, Sandra Day O'Conner." Brainy Quote at https://www.brainyquote.com/topics/sanctions-quotes.

Additionally, the chapter explores strategies for businesses to avoid engaging with sanctioned entities or individuals, thereby mitigating the risk of exposure to legal and reputational risks. Through rigorous screening processes, enhanced Know Your Customer (KYC) procedures, and leveraging advanced technology solutions, businesses can identify and avoid dealing with entities or individuals subject to sanctions, thereby safeguarding their reputation and integrity. Furthermore, proactive engagement with regulatory authorities and industry stakeholders can provide businesses with valuable insights and guidance on compliance best practices in navigating complex sanctions regimes.

The chapter examines the implications of internal surveillance measures on businesses, particularly concerning the accessibility of surveillance data by government agencies. In an increasingly digitized and interconnected world, businesses must grapple with the challenges of balancing privacy concerns with regulatory requirements for data retention and disclosure. By implementing robust data governance policies, encryption technologies, and access controls, businesses can protect sensitive information from unauthorized access and mitigate the risk of government intrusion into their internal surveillance systems.

Furthermore, the chapter demarcates the nuances of government surveillance practices and how they vary across jurisdictions, posing unique challenges and considerations for businesses operating globally. From differences in legal frameworks and oversight mechanisms to cultural norms and political dynamics, businesses must navigate a complex landscape of regulatory requirements and expectations regarding government surveillance. By conducting comprehensive risk assessments and tailoring their compliance efforts to specific jurisdictions, businesses can adapt to local regulations while upholding global standards of privacy and data protection.

Chapter 36 provides a comprehensive examination of the challenges and implications of sanctions and surveillance for businesses operating in today's global marketplace. By proactively preparing for sanctions, avoiding dealings with sanctioned entities, safeguarding internal surveillance data, and navigating the complexities of government surveillance across jurisdictions, businesses can enhance their resilience and mitigate the risks associated with regulatory compliance. Through a combination of strategic foresight, robust compliance frameworks, and proactive engagement with regulators and industry stakeholders, businesses can effectively navigate the intricate landscape of sanctions and surveillance while safeguarding their interests and reputation in an ever-changing regulatory environment.

Avoiding Becoming Sanctioned

Avoiding sanctions requires a comprehensive understanding of the regulatory landscape and a proactive approach to compliance. Businesses can take several measures to minimize the risk of being sanctioned by governments or international bodies. First, conducting thorough due diligence on potential business partners, customers, and suppliers is essential to ensure compliance with sanctions laws and regulations.[649] This includes screening parties against relevant sanctions lists and conducting enhanced Know Your Customer (KYC) checks to verify their identities and backgrounds.

---

[649] DeDominicis, B.E., 2022. Web3, Hegemony And Anonymity: The Transnational Social Identity Dynamics Of Globalized National Political Economic Interdependency. *Global Journal of Business Research*, *16*(1), pp.91-119.

Second, implementing robust compliance programs and risk assessment frameworks is crucial for identifying and mitigating potential sanctions risks. Businesses should establish clear policies and procedures for screening transactions and business relationships, as well as mechanisms for monitoring and reporting suspicious activities.[650] By integrating sanctions compliance into their overall risk management strategy, businesses can effectively identify and address potential vulnerabilities before they escalate into compliance breaches.

Next, maintaining open channels of communication with regulatory authorities and seeking guidance on compliance obligations can help businesses navigate the complexities of sanctions regimes.[651] Proactively engaging with regulators, attending industry seminars, and participating in compliance forums can provide valuable insights and updates on regulatory developments, enabling businesses to stay ahead of evolving sanctions risks and adjust their compliance efforts accordingly.

Lastly, investing in employee training and awareness programs is essential to ensure that staff members are well-informed about sanctions laws and regulations and understand their roles and responsibilities in maintaining compliance. By educating employees about the implications of sanctions violations, providing regular updates on sanctions developments, and fostering a culture of compliance and integrity, businesses can empower their workforce to identify and address potential sanctions risks effectively. Through these proactive measures, businesses can minimize the likelihood of being sanctioned and maintain their reputation and integrity in the marketplace.

---

[650] *Id.*
[651] *Id.*

Avoiding dealing with Sanctioned Companies

Avoiding working with sanctioned companies is crucial for businesses to uphold their integrity, protect their reputation, and ensure compliance with international laws and regulations. Collaborating with sanctioned entities can expose businesses to significant legal, financial, and reputational risks, including hefty fines, legal penalties, and damage to brand reputation. Therefore, it is imperative for businesses to exercise due diligence and vigilance when engaging in business relationships to avoid inadvertently dealing with sanctioned entities.

One primary reason for avoiding working with sanctioned companies is to comply with regulatory requirements and international sanctions regimes. Governments and international bodies impose sanctions on entities or individuals deemed to pose a threat to national security, international peace, or human rights. By adhering to sanctions laws and regulations and avoiding transactions with sanctioned entities, businesses can demonstrate their commitment to ethical conduct, legal compliance, and corporate responsibility.[652]

Furthermore, collaborating with sanctioned companies can tarnish a business's reputation and credibility in the marketplace. Customers, investors, and other stakeholders may view associations with sanctioned entities as a sign of poor judgment, ethical lapses, or disregard for legal and regulatory obligations. Consequently, businesses risk losing trust, credibility, and market share, as well as facing reputational damage that can have long-lasting consequences for their brand and business relationships.

---

[652] United Nations Model Law on Cryptocurrency and Digital Assets, Draft (2024).

Engaging with sanctioned companies can result in severe financial consequences for businesses, including fines, legal costs, and loss of revenue. Governments and regulatory authorities impose hefty penalties on businesses found to have violated sanctions laws or engaged in prohibited transactions with sanctioned entities. These penalties can have a significant impact on a business's financial health, profitability, and ability to operate effectively in the marketplace. Therefore, avoiding working with sanctioned companies is not only a matter of legal compliance but also a prudent risk management strategy to safeguard the financial interests and sustainability of the business.

Internal Surveillance

The need for internal security and surveillance within businesses is paramount to safeguarding sensitive information, preventing unauthorized access, and detecting potential security threats. By implementing robust security measures such as access controls, encryption, and intrusion detection systems, businesses can protect their data assets and infrastructure from cyber threats, insider attacks, and other security breaches.[653] Internal surveillance plays a crucial role in monitoring employee activities, detecting suspicious behavior, and ensuring compliance with company policies and procedures.

However, while internal surveillance can enhance security and accountability within organizations, it also raises concerns about privacy and data protection. Employees have a reasonable expectation of privacy in the workplace,

---

[653] Harsono, H., 2022. The utilization of WEB3 native resources to create a centralized base of authoritarian power. *Journal of International Affairs*, *75*(1), pp.153-168.

and businesses must strike a balance between monitoring activities for security purposes and respecting the privacy rights of their staff. Transparent communication, clear policies, and consent mechanisms are essential to ensure that employees understand the purpose and scope of surveillance activities and consent to their use in compliance with applicable laws and regulations.

Furthermore, businesses must be mindful of the potential for government agencies to seize internal surveillance files during investigations or legal proceedings. Governments may compel businesses to disclose surveillance data or provide access to their surveillance systems as part of criminal investigations, regulatory inquiries, or national security matters.[654] In such cases, businesses must comply with lawful requests for information while also protecting the privacy rights of individuals and ensuring that surveillance data is handled and disclosed in accordance with applicable laws and regulations.

To comply with the law and protect the privacy of their customers and staff, businesses should adopt several best practices. First, businesses should implement clear and transparent policies governing the collection, use, and retention of surveillance data, ensuring that employees are aware of their rights and responsibilities regarding privacy and data protection.[655] Additionally, businesses should conduct privacy impact assessments and risk assessments to evaluate the potential privacy implications of their surveillance activities and implement appropriate safeguards to mitigate risks.

---

[654] *Id*

[655] Wu, *supra*.

Second, businesses should implement technical and organizational measures to protect the security and integrity of surveillance data, including encryption, access controls, and data anonymization techniques. By encrypting sensitive data, limiting access to authorized personnel, and anonymizing personally identifiable information, businesses can reduce the risk of unauthorized access or misuse of surveillance data and enhance the privacy and security of their systems.[656]

Third, businesses should establish clear procedures and protocols for responding to government requests for surveillance data, ensuring that requests are lawful, proportionate, and justified. Businesses should seek legal advice and consult with privacy experts to assess the validity of government requests and determine the appropriate course of action to protect the rights and interests of their customers and staff.

Businesses should regularly review and audit their surveillance practices to ensure compliance with applicable laws and regulations and identify any areas for improvement. By conducting internal audits, training sessions, and compliance assessments, businesses can ensure that their surveillance activities are conducted in a lawful, ethical, and responsible manner, respecting the privacy rights of individuals and maintaining the trust and confidence of their stakeholders.[657]

---

[656] Ma, W. and Huang, K., 2022. *Blockchain and Web3: Building the cryptocurrency, privacy, and security foundations of the metaverse.* John Wiley & Sons.

[657] DeDominicis, B.E., 2022. Web3, Hegemony And Anonymity: The Transnational Social Identity Dynamics Of Globalized National Political Economic Interdependency. *Global Journal of Business Research*, *16*(1), pp.91-119.

While internal security and surveillance are essential for protecting business assets and detecting security threats, businesses must also uphold the privacy rights of their customers and staff and comply with applicable laws and regulations governing surveillance activities. By implementing transparent policies, robust security measures, and best practices for data protection, businesses can strike a balance between security and privacy, ensuring that their surveillance activities are conducted responsibly and ethically while safeguarding the rights and interests of all stakeholders involved.

Government Surveillance

Government surveillance refers to the monitoring, interception, and collection of information by government agencies for various purposes, including national security, law enforcement, and intelligence gathering. Governments employ a range of methodologies and techniques to conduct surveillance activities, often leveraging advanced technologies and legal authorities to collect data on individuals, organizations, and communications networks. While government surveillance can play a crucial role in protecting public safety and national security, it also raises concerns about privacy, civil liberties, and the potential for abuse of power.

Common methodologies used by governments for surveillance include:

- Signals intelligence (SIGINT): Governments intercept and analyze electronic communications, such as phone calls, emails, and internet traffic, to gather intelligence on potential threats and targets.[658]

[658] Ball, D. and Windrem, R., 1989. Soviet signals intelligence

- Closed-circuit television (CCTV) surveillance: Governments deploy surveillance cameras in public spaces, transportation hubs, and critical infrastructure to monitor activities, deter criminal behavior, and enhance public safety.[659]
- Covert surveillance: Governments conduct undercover operations, stakeouts, and physical surveillance to gather evidence, monitor suspects, and gather intelligence on criminal or terrorist activities.[660]
- Data mining and analysis: Governments collect and analyze large volumes of data from various sources, including public records, social media, and financial transactions, to identify patterns, trends, and potential threats.[661]
- Metadata analysis: Governments collect and analyze metadata, such as call records, location data, and internet browsing history, to track individuals' movements, associations, and activities.[662]
- Biometric surveillance: Governments use biometric technologies, such as facial recognition, fingerprint scanning, and iris scanning, to identify and track individuals in public spaces, airports, and border crossings.[663]

(SIGINT): organization and management. *Intelligence and National Security*, *4*(4), pp.621-659.

[659] Hartmus, D.M., 2014. Government guidelines for CCTV: A comparison of four countries. *International Journal of Public Administration*, *37*(6), pp.329-338.

[660] Slobogin, C., 2014. Standing and covert surveillance. *Pepp. L. Rev.*, *42*, p.517.

[661] Cate, F.H., 2008. Government data mining: The need for a legal framework. *Harv. CR-CLL Rev.*, *43*, p.435.

[662] Quarati, A., 2023. Open government data: usage trends and metadata quality. *Journal of Information Science*, *49*(4), pp.887-910.

[663] Al Rousan, M. and Intrigila, B., 2020. A comparative analysis of biometrics types: literature review. *Journal of Computer Science*,

- Mass surveillance programs: Governments deploy bulk collection programs to collect and store vast amounts of data on citizens' communications, activities, and behaviors, often without individualized suspicion or warrants.[664]

To comply with the law while maintaining the right to privacy, companies can adopt several methods and best practices:

1. Transparency and accountability: Companies should be transparent about their data collection and surveillance practices, clearly communicate their privacy policies to customers and employees, and establish mechanisms for accountability and oversight to ensure compliance with applicable laws and regulations.
2. Minimization of data collection: Companies should collect only the data necessary for legitimate business purposes and minimize the collection of sensitive or personally identifiable information that is not essential to their operations. By adopting a data minimization approach, companies can reduce the risk of privacy violations and unauthorized access to sensitive data.
3. Consent and user control: Companies should obtain explicit consent from individuals before collecting, processing, or sharing their personal data, and provide users with meaningful choices and controls over how their data is used. By empowering users to exercise control over their privacy settings and

---

16(12), pp.1778-1788.

[664] Ram, N. and Gray, D., 2020. Mass surveillance in the age of COVID-19. *Journal of Law and the Biosciences*, *7*(1), p.lsaa023.

preferences, companies can build trust and confidence in their products and services.

4. Encryption and data security: Companies should implement robust encryption and data security measures to protect sensitive information from unauthorized access, interception, or tampering. By encrypting data in transit and at rest, implementing access controls, and regularly auditing their security controls, companies can mitigate the risk of data breaches and unauthorized disclosure of sensitive information.
5. Legal compliance and due process: Companies should comply with applicable laws, regulations, and legal standards governing surveillance activities, including obtaining warrants and court orders where required, respecting individuals' rights to due process and judicial oversight, and challenging government requests for data that are overly broad or lack legal basis.
6. Privacy by design: Companies should adopt a privacy-by-design approach to product and system development, integrating privacy principles and controls into the design, architecture, and functionality of their products and services from the outset. By embedding privacy into the design process, companies can proactively address privacy risks and minimize the likelihood of privacy violations down the line.
7. Transparency reports and accountability mechanisms: Companies should publish transparency reports detailing government requests for user data, surveillance activities, and compliance with legal and regulatory requirements. By providing transparency into their surveillance practices and accountability mechanisms, companies can enhance trust, transparency, and

accountability in their interactions with users, regulators, and other stakeholders.

8. Legal challenges and advocacy: Companies should advocate for legal reforms and protections to safeguard privacy rights and civil liberties in the face of government surveillance activities. By engaging in legal challenges, lobbying efforts, and public advocacy campaigns, companies can support efforts to strengthen privacy laws, enhance transparency and accountability in government surveillance, and protect individuals' rights to privacy and due process.

While government surveillance is a necessary tool for ensuring national security and public safety, it also presents significant challenges to individual privacy, civil liberties, and corporate accountability. By adopting transparent and accountable surveillance practices, minimizing data collection, obtaining user consent, and implementing robust security measures, companies can navigate the complex regulatory landscape and protect the privacy rights of their customers and employees. Moreover, by advocating for legal reforms, promoting transparency and accountability, and challenging government overreach, companies can contribute to the ongoing dialogue on privacy rights and civil liberties in the digital age. As governments continue to expand their surveillance capabilities and authorities, it is essential for companies to uphold the principles of privacy, transparency, and accountability in their surveillance activities, ensuring that they comply with the law while safeguarding the rights and interests of all stakeholders involved.

Conclusion

This chapter has provided a comprehensive exploration of the intricate landscape of sanctions, surveillance, and the challenges they pose for businesses operating in today's globalized world. The chapter has shed light on the complexities of navigating international sanctions regimes, avoiding dealings with sanctioned entities, and addressing the implications of government surveillance on privacy rights and civil liberties. As businesses strive to uphold ethical standards, comply with legal requirements, and protect their interests in an increasingly regulated environment, they must adopt proactive measures to mitigate the risks associated with sanctions and surveillance while maintaining their commitment to transparency, accountability, and corporate responsibility.

The parts of this chapter have underscored the importance of robust compliance programs, risk assessment frameworks, and due diligence measures in mitigating the risks of sanctions violations and avoiding entanglements with sanctioned entities. By implementing clear policies, conducting thorough screenings, and fostering a culture of compliance, businesses can enhance their resilience to regulatory risks and protect their reputation and integrity in the marketplace. Additionally, the chapter has highlighted the need for businesses to remain vigilant and adaptable in the face of evolving sanctions regimes and regulatory requirements, proactively monitoring developments and adjusting their compliance efforts accordingly.

Furthermore, the chapter has emphasized the critical role of transparency, accountability, and stakeholder engagement in addressing the challenges of government surveillance and protecting individual privacy rights. By advocating for legal reforms, promoting transparency reports, and challenging government overreach, businesses can contribute to the ongoing dialogue on privacy rights and civil liberties in the digital age. Moreover, by adopting privacy-by-design principles, implementing robust security measures, and respecting user consent, businesses can build trust, enhance transparency, and protect the privacy rights of their customers and employees.

As businesses navigate the complexities of sanctions and surveillance in an interconnected world, they must strike a delicate balance between compliance with legal and regulatory requirements and upholding ethical standards and corporate values. By embracing transparency, accountability, and responsible business practices, businesses can navigate the challenges of sanctions and surveillance with integrity and resilience, safeguarding their interests and contributing to a more transparent, accountable, and ethical business environment.

Chapter 37: Financial Stability, Systematic Risk, and Access to Systems

"Arguing that you don't care about your right to privacy because you have nothing to hide is no different than saying you don't care about free speech because you have nothing to say."

Edward Snowden[665]

Chapter 37 completes our analysis with the critical intersections of financial stability, systematic risk, and access control within the realm of business security. In today's digital age, where data serves as a cornerstone of business operations, ensuring the security and stability of financial systems is paramount to the success and resilience of companies. This chapter explores the multifaceted nature of financial stability and its intrinsic connection to the security of data assets, highlighting the importance of robust security measures in mitigating systematic risks and preserving the integrity of financial systems.

One key aspect addressed in this chapter is the significance of data security in maintaining financial stability. As businesses increasingly rely on digital platforms and technologies to conduct transactions, manage assets, and interact with customers, safeguarding sensitive financial data from unauthorized access, cyber threats, and data breaches becomes imperative. By implementing rigorous data protection measures, encryption protocols, and access controls, businesses can enhance the security of their financial systems and mitigate the risk of financial fraud, identity theft, and regulatory non-compliance.

[665] (2024). "Cybersecurity Quotes, Edward Snowden." Pinngle at https://pinngle.me/blog/99-inspiring-cybersecurity-quotes/index.html.

Furthermore, the chapter examines the need for tailored security approaches across different fields and industries. Notably, while the principles of data security remain universal, the specific security requirements and challenges vary depending on the nature of the business, industry regulations, and technological infrastructure. By understanding the unique security needs and risk profiles of their respective fields, businesses can implement customized security solutions that address specific threats and vulnerabilities, thereby enhancing the resilience and effectiveness of their financial systems.

The chapter explores the role of artificial intelligence (AI) in augmenting corporate planning and risk management strategies to decrease systematic risk. AI-powered analytics and predictive modeling techniques enable businesses to analyze vast amounts of data, identify patterns, and forecast potential risks with greater accuracy and efficiency. By leveraging AI-driven insights, businesses can make data-driven decisions, optimize resource allocation, and proactively mitigate systemic risks, thereby enhancing the stability and resilience of financial systems in an increasingly dynamic and interconnected environment.

Lastly, the chapter underscores the importance of controlling access to financial systems and sensitive data to prevent unauthorized activities and mitigate insider threats. By implementing robust access control mechanisms, authentication protocols, and user monitoring systems, businesses can enforce strict access policies, limit privileges to authorized personnel, and detect suspicious activities in real-time. By maintaining granular control over access permissions and implementing multi-factor authentication, businesses can enhance the security posture of their financial systems and minimize the risk of data breaches and financial fraud.

Financial Stability and the Security of Your Data

Financial stability plays a pivotal role in enhancing the security and validity of a company's data assets by providing a solid foundation for business operations and risk management. A financially stable company has the resources and capacity to invest in robust data infrastructure, security protocols, and risk mitigation strategies, thereby fortifying its defenses against cyber threats, data breaches, and other security vulnerabilities.[666] By allocating sufficient resources to data security initiatives and compliance efforts, financially stable companies can implement comprehensive security measures that protect sensitive data, uphold regulatory requirements, and preserve the integrity of their data assets.

Financial stability enables companies to adopt best practices and industry standards for data management and security, thereby enhancing the accuracy, reliability, and validity of their data. With adequate financial resources and organizational support, companies can invest in state-of-the-art data management systems, data governance frameworks, and data quality controls that ensure the integrity and consistency of their data assets.[667] By implementing standardized data management practices and quality assurance processes, financially stable companies can minimize errors, inconsistencies, and discrepancies in their data, thereby enhancing the security and trustworthiness of their data assets.

---

[666] Onyshchuk, S.V., Onyshchuk, I.I., Petroye, O. and Chernysh, R., 2020. Financial stability and its impact on national security state: Organizational and legal aspects.
[667] Heath, M.R.M. and Goksu, E.B., 2017. *Financial stability analysis: What are the data needs?*. International Monetary Fund.

Furthermore, financial stability empowers companies to invest in advanced technologies and cybersecurity solutions that enhance the security and resilience of their data infrastructure. With adequate financial resources and technical expertise, companies can deploy cutting-edge encryption techniques, intrusion detection systems, and threat intelligence platforms that safeguard sensitive data from unauthorized access, cyber attacks, and data breaches.[668] By leveraging innovative technologies and security solutions, financially stable companies can stay ahead of emerging threats, detect security incidents in real-time, and respond swiftly to mitigate risks and minimize potential damages to their data assets.

Additionally, financial stability enables companies to implement robust risk management strategies and contingency plans that mitigate the impact of data security incidents and ensure business continuity. With adequate financial reserves and risk management frameworks in place, companies can conduct comprehensive risk assessments, identify potential vulnerabilities and threats to their data assets, and develop proactive mitigation measures and response plans to address security incidents effectively. By investing in risk management capabilities and resilience-building initiatives, financially stable companies can minimize the likelihood of data security breaches, mitigate the financial and reputational consequences of security incidents, and maintain the stability and validity of their data assets.

Financial stability enhances the credibility and trustworthiness of a company's data assets in the eyes of customers, partners, and regulators. A financially stable company demonstrates its commitment to data security,

---

[668] Onyshchuk et al, *supra*.

compliance, and risk management, thereby instilling confidence in stakeholders and enhancing the perceived reliability and integrity of its data assets. By maintaining financial stability and investing in data security initiatives, companies can strengthen their competitive advantage, attract and retain customers, and build long-term relationships based on trust, transparency, and accountability.

Financial stability is a critical enabler of data security and integrity, providing companies with the resources, capabilities, and resilience needed to protect their data assets from security threats, ensure compliance with regulatory requirements, and preserve the trust and confidence of stakeholders. By prioritizing financial stability and investing in data security initiatives, companies can enhance the stability, validity, and security of their data assets, thereby positioning themselves for long-term success and sustainability in an increasingly digital and interconnected world.

Different Security for Different Areas

Different fields indeed entail varying security requirements, and within a company, diverse areas necessitate tailored security measures to address their unique vulnerabilities and safeguard sensitive assets effectively. For instance, industries like healthcare and finance handle highly sensitive personal and financial data, demanding stringent security protocols to adhere to regulatory mandates and protect against data breaches.[669] In contrast, sectors such as manufacturing may prioritize physical security to safeguard industrial machinery and intellectual property, requiring

[669] Gensler, G. and Bailey, L., 2020. Deep learning and financial stability. *Available at SSRN 3723132.*

robust access controls and surveillance systems. By recognizing these variances, companies can deploy specialized security solutions tailored to the specific needs and risk profiles of each field, ensuring comprehensive protection across all aspects of their operations.

Even within a company, different departments or divisions may warrant distinct security strategies based on their roles, responsibilities, and levels of data access. For example, the finance department may necessitate heightened access controls and encryption measures to safeguard financial transactions and sensitive customer information, while the marketing department may prioritize cybersecurity awareness training and phishing prevention to mitigate the risk of social engineering attacks. By customizing security protocols to align with the unique requirements and operational priorities of each department, companies can bolster their overall security posture and minimize the likelihood of security breaches or data compromises.

Furthermore, varying security requirements may stem from differences in technological infrastructures, data storage systems, and communication networks utilized across different fields or departments within a company. For instance, research and development teams may rely on cloud-based platforms and collaborative tools to facilitate innovation and collaboration, necessitating robust encryption and authentication mechanisms to protect intellectual property and confidential research data. Meanwhile, customer service departments may utilize customer relationship management (CRM) systems to manage client interactions and inquiries, requiring secure authentication and access controls to safeguard customer information and maintain privacy.

Additionally, regulatory compliance obligations and industry standards may influence the security requirements imposed on different fields or departments within a company. Industries subject to stringent regulations, such as healthcare (HIPAA), finance (PCI-DSS), or data privacy (GDPR), must adhere to specific security standards and compliance frameworks to mitigate legal risks and regulatory penalties.[670] Consequently, companies must implement tailored security measures and compliance initiatives tailored to meet the regulatory requirements of each field or department, ensuring adherence to applicable laws and regulations while maintaining data integrity and confidentiality.

In conclusion, recognizing the diverse security requirements inherent in different fields and departments is essential for companies to develop comprehensive security strategies that address the unique challenges and vulnerabilities present in each area of their operations. By tailoring security measures to align with specific industry regulations, operational priorities, and technological infrastructures, companies can enhance their overall security posture and effectively mitigate risks across all aspects of their business. Furthermore, fostering a culture of security awareness and proactive risk management enables companies to adapt and evolve their security strategies in response to emerging threats and changing regulatory landscapes, ensuring ongoing protection of sensitive assets and data.

Preventing Systematic Risk

Companies operating in the web3 space can enhance their security posture by implementing proactive measures to

---

670 *Id.*

prevent systematic risks, thereby safeguarding their operations, assets, and stakeholders from potential vulnerabilities and disruptions. One key strategy for mitigating systematic risk involves conducting comprehensive risk assessments to identify and evaluate potential threats, vulnerabilities, and dependencies within the organization's ecosystem.[671] By systematically analyzing internal and external factors that could impact the company's operations, such as technology dependencies, regulatory changes, and market volatility, companies can proactively identify areas of weakness and develop targeted mitigation strategies to address systemic risks before they escalate into larger threats.

Furthermore, leveraging advanced technologies and artificial intelligence (AI) can enable companies to augment their corporate planning and risk management efforts, thereby reducing systematic risk and enhancing overall security. AI-powered analytics and predictive modeling techniques empower companies to analyze vast amounts of data, identify patterns, and forecast potential risks with greater accuracy and efficiency.[672] By leveraging AI-driven insights, companies can make data-driven decisions, optimize resource allocation, and proactively mitigate systemic risks, thereby enhancing the stability and resilience of their operations in an increasingly dynamic and interconnected environment.

Fostering a culture of collaboration, transparency, and accountability across all levels of the organization can help companies effectively manage systematic risks and enhance security.[673] By promoting open communication channels,

---

[671] Carmen, T.O.D.E.R.A.□.C.U. and Gabriel, N.V., THE POTENTIAL OF WEB3 REGARDING DECENTRALIZED FINANCE-DEFI.
[672] *Id.*

sharing information and insights, and encouraging cross-functional collaboration, companies can leverage collective expertise and insights to identify, assess, and address systemic risks more effectively. Additionally, establishing clear accountability mechanisms and governance structures ensures that stakeholders understand their roles and responsibilities in managing systemic risks and complying with security protocols, thereby enhancing overall organizational resilience and security.

Furthermore, diversifying and decentralizing operations and infrastructure can help companies mitigate systemic risks by reducing reliance on single points of failure and enhancing redundancy and resilience. By distributing resources, data, and operations across multiple locations, networks, and providers, companies can minimize the impact of potential disruptions and vulnerabilities, thereby enhancing the security and stability of their operations. Additionally, leveraging decentralized technologies, such as blockchain, can further enhance security by providing tamper-proof and transparent transaction records, reducing the risk of data manipulation, fraud, and cyber attacks.

Preventing systematic risks requires a multifaceted approach that encompasses proactive risk management, technological innovation, organizational culture, and operational resilience.[674] By conducting comprehensive risk assessments, leveraging advanced technologies, fostering collaboration and transparency, and diversifying operations and infrastructure, companies in the web3 space can enhance their security posture and mitigate the impact of

---

[673] Nwogugu, M.C., 2022. ESG, Inequality And International Political Economy Problems Inherent In Regulation Of DeFi, Web5/Web3, CryptoCurrencies And Fractional-NFTs. *Web5/Web3, CryptoCurrencies And Fractional-NFTs (Revised 2022).*
[674] *Id.*

potential threats and vulnerabilities. By adopting a proactive and strategic approach to risk management, companies can strengthen their resilience, adaptability, and competitiveness in an ever-evolving digital landscape.

Controlling Access to Systems

Security, fundamentally, revolves around controlling access to systems and ensuring that individuals with access are authorized for legitimate purposes, thus safeguarding the integrity, confidentiality, and availability of sensitive assets and information. Access control mechanisms serve as the first line of defense against unauthorized access, malicious activities, and insider threats, thereby forming a critical component of an organization's overall security posture. By implementing robust access control measures, such as user authentication, authorization policies, and encryption protocols, companies can restrict access to sensitive data and resources to only those individuals with a legitimate need, thereby mitigating the risk of unauthorized access and potential security breaches.

Access control encompasses various layers of security, including physical, logical, and administrative controls, to enforce strict access policies and prevent unauthorized entry or activities within the system. Physical access controls, such as biometric authentication, access badges, and surveillance systems, restrict physical access to facilities and sensitive equipment, thereby preventing unauthorized individuals from gaining physical access to critical infrastructure.[675] Logical access controls, on the other hand, govern access to digital resources, networks,

[675] Ding, W., Hou, J., Li, J., Guo, C., Qin, J., Kozma, R. and Wang, F.Y., 2022. DeSci based on Web3 and DAO: A comprehensive overview and reference model. *IEEE Transactions on Computational Social Systems*, *9*(5), pp.1563-1573.

and applications, ensuring that only authorized users can access and interact with sensitive data and systems. By implementing robust authentication mechanisms, role-based access controls, and encryption protocols, companies can enforce granular access controls and prevent unauthorized access to sensitive information.

Furthermore, ensuring that individuals with access privileges are authorized for legitimate actions entails implementing comprehensive identity and access management (IAM) frameworks and security policies. IAM solutions enable companies to centrally manage user identities, access permissions, and authentication mechanisms, thereby streamlining access control processes and ensuring that users have appropriate access privileges based on their roles, responsibilities, and organizational needs. By implementing IAM solutions, companies can enforce strong authentication requirements, enforce least privilege access principles, and monitor user activities in real-time to detect and prevent unauthorized access or suspicious behavior.

Additionally, access control encompasses proactive measures to monitor and audit user activities, detect anomalous behavior, and respond swiftly to potential security incidents or breaches. By implementing robust logging mechanisms, intrusion detection systems, and security analytics tools, companies can monitor user activities, track access patterns, and identify potential security threats or anomalies in real-time. Moreover, conducting regular security audits, penetration testing, and vulnerability assessments enables companies to identify weaknesses and gaps in their access control mechanisms, thereby proactively addressing security vulnerabilities and strengthening their overall security posture.

Controlling access to systems and ensuring that authorized users are there for legitimate actions is a fundamental principle of security that underpins the integrity, confidentiality, and availability of sensitive assets and information. By implementing robust access control mechanisms, enforcing strong authentication and authorization policies, and monitoring user activities in real-time, companies can mitigate the risk of unauthorized access, prevent security breaches, and safeguard their critical infrastructure and data assets.[676] Additionally, fostering a culture of security awareness, accountability, and compliance ensures that employees understand their roles and responsibilities in protecting sensitive information and adhering to security policies, thereby enhancing overall security and resilience against evolving threats and vulnerabilities.

Conclusion

Chapter 37 has delved into the multifaceted realm of financial stability, systematic risk management, and access control within the context of security in the web3 space. It has emphasized the critical importance of financial stability as a cornerstone of effective risk management and security, highlighting how a robust financial foundation enables companies to invest in comprehensive security measures, technology, and infrastructure to protect their operations and data assets. The chapter has underscored the significance of proactive risk assessment, technological innovation, and organizational resilience in mitigating systematic risks and enhancing overall security posture.

---

[676] Voshmgir, S., 2020. *Token economy: How the Web3 reinvents the internet* (Vol. 2). Token Kitchen.

Furthermore, the chapter has highlighted the role of advanced technologies, such as artificial intelligence and decentralized systems, in augmenting corporate planning, risk management, and security efforts. By leveraging AI-driven analytics, predictive modeling, and blockchain technology, companies can gain deeper insights into potential risks, enhance decision-making processes, and strengthen the security and resilience of their operations in an increasingly dynamic and interconnected environment. Additionally, the chapter has emphasized the importance of fostering a culture of collaboration, transparency, and accountability to effectively manage systematic risks and enhance security across all levels of the organization.

The chapter has underscored the significance of access control as a fundamental principle of security, emphasizing the importance of controlling access to systems and ensuring that individuals with access privileges are authorized for legitimate actions. By implementing robust access control mechanisms, identity and access management solutions, and monitoring and auditing processes, companies can mitigate the risk of unauthorized access, prevent security breaches, and safeguard their critical infrastructure and data assets. Furthermore, the chapter has highlighted the importance of aligning security strategies with regulatory requirements, industry standards, and best practices to ensure compliance and mitigate legal and reputational risks.

Chapter 37 has provided valuable insights into the interconnected nature of financial stability, systematic risk management, and access control within the realm of security in the web3 space. By adopting a proactive and strategic approach to risk management, leveraging advanced technologies, and fostering a culture of security awareness and accountability, companies can enhance their security posture, mitigate risks, and strengthen their resilience in an ever-evolving digital landscape. Moving forward, it is imperative for companies to prioritize security as a strategic imperative and invest in comprehensive security measures and initiatives to protect their operations, data assets, and stakeholders from emerging threats and vulnerabilities.

Conclusion of Part VIII

In concluding Part VIII, a comprehensive understanding of security in the web3 landscape emerges, emphasizing the multifaceted challenges and proactive strategies essential for safeguarding digital environments. Throughout these chapters, crucial aspects of investment monitoring, regulatory compliance, cybercrime mitigation, and access control have been meticulously explored, shedding light on the intricacies of securing operations in a rapidly evolving digital era.

Chapter 34 looked into the critical realm of investment and development monitoring within the context of AI and machine learning security, highlighting the importance of robust mechanisms to track investment performance, assess developmental progress, and ensure supply chain integrity. It underscored the significance of continuous monitoring, portfolio analysis, and developmental planning in driving innovation and managing projects effectively.

Following this, Chapter 35 elucidated the intricate landscape of cross-border cybercrime, encompassing data breaches, ransomware attacks, supply chain vulnerabilities, and espionage threats. It emphasized the imperative for companies to bolster their cybersecurity defenses, implement proactive measures, and foster a culture of security awareness to mitigate the risks posed by cyber threats and state-sponsored attacks.

In Chapter 36, the focus shifted towards the impact of government actions, including sanctions and surveillance, on legitimate business activities. It underscored the importance of compliance with regulatory requirements, proactive risk management, and safeguarding internal data against government surveillance while upholding privacy rights and ensuring transparency in business operations.

Last, Chapter 37 underscored the symbiotic relationship between financial stability, systematic risk management, and access control in shaping the security landscape of web3 companies. It emphasized the significance of financial resilience, proactive risk assessment, advanced technologies, and robust access control mechanisms in mitigating threats, enhancing security posture, and fostering organizational resilience.

Overall, Part VIII has provided invaluable insights into the complex interplay of factors influencing security in the web3 environment, ranging from investment strategies and regulatory compliance to cybercrime mitigation and access control. It has underscored the importance of proactive risk management, technological innovation, regulatory compliance, and collaboration in safeguarding digital assets, protecting stakeholders, and ensuring the integrity and resilience of web3 companies in an increasingly interconnected and dynamic landscape.

Conclusion of Book

If you had to talk to your younger self, would you imagine that you were telling them that you just read a book on digital security. Ignore the fact that the level of technology today falls somewhere between that which your child-self would not believe could happen and that humanity has failed its mandate of being a utopia by the third decade of the new millennia- would you believe that corporate security is something that you are involved in? This is where we are as a society, where digital security has become something that every person needs to take note of.

As we conclude this book, take control of your business life by taking control of your security. If you set back from security, you are allowing someone else to have the helm of your destiny. This is not to say that you should not hire consultants, security experts, or staff to work security- it is merely to say that you have a duty to your company to ensure that you are an active part of the security. Whether worker or owner, custodian or secretary, planner or implementor- you are someone whom your company relies upon for continued security. Take the role seriously.

To this end, we end. I hope that this text has been useful to you for helping you understand how the world is changing. Digital security, by and large, is a bigger industry than physical security in the modern world. You are a part of it. Take your part seriously and business can be an area of success and certainty for you. Take this part lightly, and this book may be in hock for you to make payments. There are few things you can control in your life, how seriously you take security is one of them.

Made in the USA
Monee, IL
04 November 2026

67381812R10282